Targeted Genome Editing Using Site-Specific Nucleases

Takashi Yamamoto
Editor

Targeted Genome Editing Using Site-Specific Nucleases

ZFNs, TALENs, and the CRISPR/Cas9 System

Springer

Editor
Takashi Yamamoto
Department of Mathematical and Life Sciences
Hiroshima University
Higashi-Hiroshima, Hiroshima, Japan

ISBN 978-4-431-55226-0 ISBN 978-4-431-55227-7 (eBook)
DOI 10.1007/978-4-431-55227-7
Springer Tokyo Heidelberg New York Dordrecht London

Library of Congress Control Number: 2014954587

Printed on acid-free paper

Springer is part of Springer Science+Business Media (www.springer.com)

Preface

Genome editing using targetable nucleases is an emerging technology for the precise genome modification of organisms ranging from bacteria to plants to animals. Its attraction is that it can be used for almost all organisms in which targeted genome modification has not been possible. To date, various modifications have been reported in cultured cells and many organisms, including gene knockout and knock-in experiments with targeting vectors, and chromosomal editing. Indeed, we are constantly surprised at the rapid progress and possibilities suggested by this technology. The development of TALENs in 2010 has enabled the genome modification of non-model organisms, while the emergence of the CRISPR/Cas9 system in 2013 has allowed us to anticipate the forthcoming new era of genome editing research.

Compared with other biotechnologies, the development of genome editing has been unusually rapid. Therefore, some researchers are unaware of its existence, while many do not fully understand the basics or how to apply the technology to their particular organism. The purpose of this book is therefore twofold. The first part sets out to introduce the history of the development of genome editing tools such as ZFN, TALEN, and CRISPR/Cas9 to beginners. The second part discusses the applications of genome editing to different organisms. However, because it is not certain how long the various methods introduced in this book will be in use, and because the applications of genome editing will undoubtedly progress, it is recommended that researchers continuously update their knowledge based on the latest genome editing technology.

In the near future, we anticipate that more precise and complex modifications such as single nucleotide substitution at multiple loci, and chromosomal editing will be possible in cultured cells and organisms. Moreover, the progress of such a convenient and safe method will undoubtedly have a high utility value in medical research. Soon, it is likely that the development of therapeutics by genome editing may no longer be just a dream. Genome editing is also hoped to be available for use in the generation of crops and livestock with useful traits. Considering these possibilities, it appears certain that genome editing will become the next generation of biotechnology.

Although the availability of genome editing is dependent on species, it is neither a difficult nor an expensive technology. Thus, I hope that young scientists will adopt this technology in their own research, and I am certain that this book will be helpful for such a purpose. Finally, I thank all authors for their contributions towards establishing this book.

Higashi-Hiroshima, Japan Takashi Yamamoto

Contents

Part I
Basics of Genome Editing

Chapter 1
Genome Editing Using Zinc-Finger Nucleases (ZFNs) and Transcription Activator-Like Effector Nucleases (TALENs)

Hiroshi Ochiai and Takashi Yamamoto

Abstract Targetable nucleases, including zinc-finger nucleases (ZFNs), transcription activator-like effector nucleases (TALENs), and clustered regularly interspaced short palindromic repeat (CRISPR)/CRISPR-associated (Cas), induce DNA double-strand breaks (DSBs) into user-defined sites. DSBs are immediately repaired through the evolutionarily conserved pathways of error-prone non-homologous end joining (NHEJ) or homology-directed repair (HDR). With the utilization of these repair processes, researchers have been able to disrupt specific genes, add exogenous DNA elements into intended genomic sites, introduce single-nucleotide substitutions, and perform many other applications. Consequently, this "genome editing" technology has revolutionized the life science field. In addition, this technology has the potential to improve agricultural products and be applicable to therapeutic use.

Here, we will introduce a brief history of targetable nuclease-mediated genome editing and the applications of the tools that the technology provides. In this chapter, we will primarily focus on ZFNs and TALENs, which are artificial proteins composed of a specific DNA-binding domain and a restriction enzyme FokI DNA-cleavage domain. We will also review the properties and construction methods of these nucleases.

Keywords DSB • Gene targeting • Genome editing • TALEN • ZFN

H. Ochiai (✉)
Research Center for the Mathematics on Chromatin Live Dynamics,
Hiroshima University, Higashi-Hiroshima 739-8530, Japan
e-mail: ochiai@hiroshima-u.ac.jp

T. Yamamoto
Department of Mathematical and Life Sciences, Graduate School of Science,
Hiroshima University, Higashi-Hiroshima 739-8526, Japan
e-mail: tybig@hiroshima-u.ac.jp

T. Yamamoto (ed.), *Targeted Genome Editing Using Site-Specific Nucleases*,
DOI 10.1007/978-4-431-55227-7_1

1.1 Introduction

With the progress of DNA sequencing technology, the genomic sequences of many organisms have now been determined. Most researchers consider the blueprint of life to be written in the genomic DNA and that the elucidation of its code, especially the functions of genes, can directly lead to an understanding of living things. To explore the functions of genes, a large number of approaches have been developed to date. Among them, homologous recombination-based gene modification is the best way to precisely analyze gene functions, but the technique is only available in restricted organisms, such as yeasts and mice. The availability of this technique strongly drives detailed studies of genomic functions in yeasts and mice, compared with other organisms. However, individual model organisms each have certain strong points for studying specific research areas. For example, compared with mice, rats are the preferred animal for disease models because of their size and simplicity of manipulation (Aitman et al. 2008), while animals with small transparent embryos, such as sea urchins, ascidians, worms, fruit flies and zebrafish, are suitable for in vivo observations of morphogenesis at single-cell resolution (Megason and Fraser 2007). Thus, to understand living things, broad knowledge from researches on a wide range of organisms is required. To overcome the unavailability of homologous recombination-based genetic modification in many organisms, two approaches are performed. The first is random integration of exogenous DNA elements that express short hairpin RNAs (shRNAs) to inhibit the production of specific protein species or overexpress some proteins to gain insights about the protein function, and the second involves treatment with chemical compounds or short inhibitory RNAs (siRNAs) to inhibit the production or function of specific protein species. However, transgenes randomly integrated into genomic DNA are sometimes not stably expressed for a long time and may disrupt the proper expression or function of endogenous genes by their integration (Emery 2011). Conversely, in most of cases, the inhibition of protein function using chemical compounds or siRNAs is transient and imperfect (Subramanya et al. 2010). Other methods, such as random mutagenesis using chemical mutagens and transposons, so-called forward genetics, are also effective for gaining insights about gene functions (Ivics et al. 2009; Yates et al. 2009). However, these approaches are extremely labor-intensive, are time-consuming to obtain mutants of interest, and involve undetectable genetic changes other than specific mutations of interest.

The emergence of genome editing technology using targetable nucleases, such as zinc-finger nucleases (ZFNs) (Urnov et al. 2010), transcription activator-like effector nucleases (TALENs) (Joung and Sander 2013), and clustered regularly interspaced short palindromic repeat (CRISPR)/ CRISPR-associated (Cas) (Mali et al. 2013a) (see Chap. 2), has changed many aspects of the life sciences field. This technology employs two key players: targetable nucleases that introduce DNA double-strand breaks (DSBs) into user-defined sites and the DSB repair mechanisms in cells (Fig. 1.1). DSBs are mainly repaired through two pathways, non-homologous end joining (NHEJ) that is inherently error-prone and homology-directed repair (HDR)

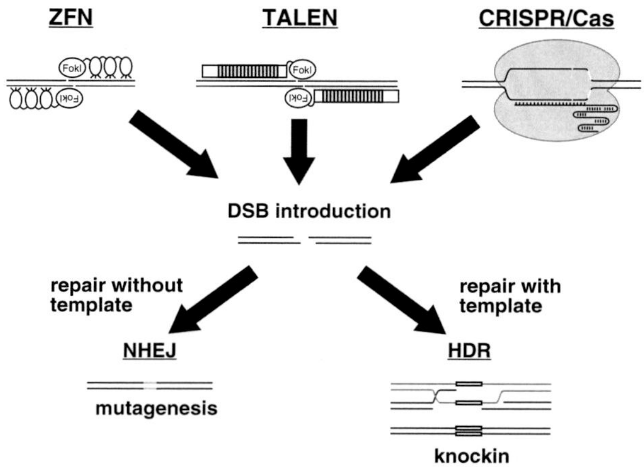

Fig. 1.1 Targetable nuclease-mediated genome editing. Targetable nucleases, such as zinc-finger nucleases (ZFNs), transcription activator-like effector nucleases (TALENs) and clustered regularly interspaced short palindromic repeat (CRISPR)/ CRISPR-associated (Cas) introduce DNA double-strand breaks (DSBs) at user-intended sites. DSBs are mainly repaired through non-homologous end joining (NHEJ) or homology-directed repair (HDR). NHEJ repair does not require a repair template and is error-prone. Therefore, when DSBs are repaired through NHEJ, small insertion or deletion mutations are frequently introduced at the sites. On the other hand, HDR requires a repair template, which could be a sister chromatid, other homologous region, or exogenous repair donor. Thus, co-introduction of targetable nucleases and exogenous repair donors facilitates transgene integration through HDR repair

that is precise compared with NHEJ but requires a repair template. Therefore, targeted DSB introduction using targetable nucleases enables efficient introduction of mutations into the site of interest (Fig. 1.1). Additionally, co-introduction of targetable nucleases and exogenous repair donors (or targeting vectors) increases the HDR-mediated targeted transgene integration (or knockin) compared with conventional gene targeting (Fig. 1.1).

To date, the genomic sequences of many model organisms have been edited by using these technologies. The model systems have involved, cultured human cells including human embryonic stem (ES) and induced pluripotent stem (iPS) cells (see Chap. 3), nematodes (see Chap. 4), silkworms (see Chap. 5), sea urchins (see Chap. 6), ascidians (see Chap. 7), fish (see Chap. 8), amphibians (see Chap. 9), mice (see Chap. 10 for mouse editing with CRISPR/Cas and Chap. 11 mouse editing with TALENs), rats (see Chap. 12), and plants (see Chap. 13).

Although the technical difficulties associated with designing ZFNs that target specific sequences of interest have hindered the expansion of ZFN-mediated genome editing, the basis of the genome editing technology was built on studies using these nucleases since 1996 (Kim et al. 1996). After the emergence of TALENs in 2010 (Christian et al. 2010) and the application of CRISPR/Cas9 for genome editing in 2013 (Cong et al. 2013; Mali et al. 2013b), which are easier to design for introduction of DSBs at the intended genomic sites than ZFNs, the technology has explosively expanded to a wide range of researchers. Consequently, we are definitely now in a genome editing era.

In this chapter, we will review the basics of the DSB repair and the molecular basis of ZFNs and TALENs. In particular, we will focus on the mechanisms of nucleotide sequence recognition and the techniques for designing sequence-specific nucleases. Finally, we will provide the examples of the genome editing using these nucleases.

1.2 DSB Repair

DSBs potentially induce loss of genetic material and oncogenic transformation through deletion and translocation mutations, and are therefore immediately repaired in cells. DSBs are mainly repaired through NHEJ or HDR (Fig. 1.1). Inherently error-prone NHEJ repair does not require a repair template and does not depend on the cell cycle. In most cases, small deletions and insertions are introduced at DSB sites through NHEJ repair. On the other hand, HDR requires a homologous repair template, such as a sister chromatid, other homologous region, or exogenous repair donor, and mainly occurs at the S/G2 phase of the cell cycle. In general, although HDR is considered to be a high-fidelity repair process, most DSBs are repaired through error-prone NHEJ. Therefore, introduction of DSBs in a gene of interest by targetable nucleases followed by NHEJ-mediated repair efficiently introduces insertion and/or deletion mutation, resulting in gene disruption (Fig. 1.1). Moreover, co-introduction of targetable nucleases and targeting vectors containing approximately 1-kb homologous regions from DSB sites facilitates HDR-mediated transgene introduction into the DSB site (Fig. 1.1). The DSB repair mechanisms are evolutionarily conserved among most of the organisms. Thus, targetable nuclease-mediated genome editing might be applicable to at least animals and plants.

1.3 ZFNs

1.3.1 Zinc-Finger Domain

ZFNs are artificial proteins comprised of a C_2H_2 zinc-finger (ZF) DNA-binding domain and a restriction enzyme FokI nuclease domain (Fig. 1.1). The C_2H_2 ZF domain is comprised of about 30 amino acids, contains two antiparallel β-sheets and an α-helix, and is characterized by a zinc ion coordinated by two cysteine residues in the β-sheets and two histidine residues in the α-helix (Fig. 1.2).

In natural proteins, the C_2H_2 ZF domain is principally involved in DNA binding. Basically, each finger recognizes a three-nucleotide subsite (target triplet, Fig. 1.2)

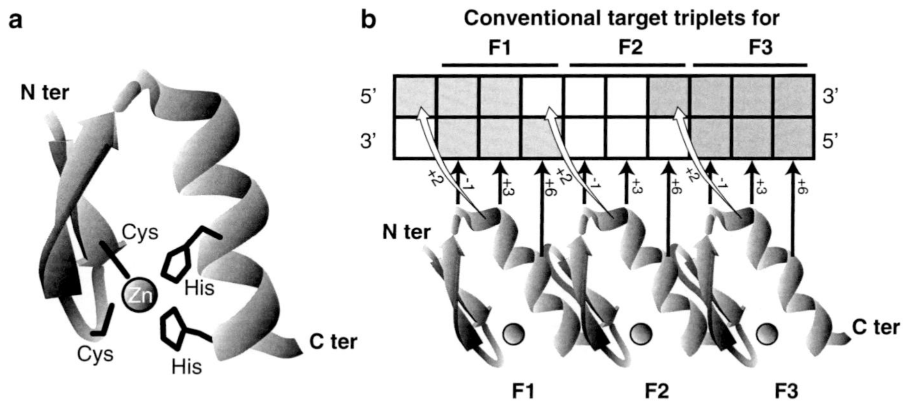

Fig. 1.2 Structure of a zinc-finger (ZF) domain. (**a**) Schematic representation of 3D structure of ZF domain. Two cysteine residues and two histidine residues coordinate a zinc ion. (**b**) Base recognition of three-finger ZFs. Fundamentally, three amino acids at helical positions −1, +3 and +6 in each ZF domain (F1, F2, and F3) directly recognize each target triplet, or subsite (*black arrows*). Sometimes, amino acid at helical position +2 is also involved in the recognition of a base in the antisense strand of an adjacent subsite (*white filled arrows*)

through its α-helix, the so-called recognition helix. It is considered that the amino acids at helical positions −1, +3 and +6 are involved in the recognition of and binding to only one strand of a target triplet. Consequently, a three-finger ZF domain recognizes a 9-bp sequence. Moreover, since ZFNs basically cleave DNA as a dimer (Figs. 1.1 and 1.3, and see below), the target of three-finger ZFNs becomes 18 bp in length, which is sufficiently long to specify a single target site in the human genome. In the helical residues, associations of specific amino acids with specific nucleotide bases are observed to some extent (Wolfe et al. 2000). However, it is considered that the base recognition is influenced by associations with other amino acids inside or outside of each finger and the nucleotide composition of the target DNA. Moreover, the residue at helical position +2 sometimes recognizes a nucleotide base in the anti-sense strand of the neighboring triplet (Fig. 1.2), resulting in an overlap between the recognition sequences of neighboring fingers. These complications make the de novo design of ZFNs, which recognize specific nucleotides, difficult. Furthermore, each finger shows strong preference for a GNN triplet for their recognition. This property strikingly restricts the target site repertoire for a pair of ZFNs (e.g. NNCNNCNNCNNNNNNGNNGNNGNN, underlines represent 6-bp spacer sequence, Fig. 1.3).

1.3.2 FokI Nuclease Domain

FokI is a type IIS restriction enzyme that cleaves DNA at a defined distance away from their recognition sites. Chandrasegaran and colleagues found that the DNA-binding and non-specific DNA-cleavage functions of FokI could be divided into

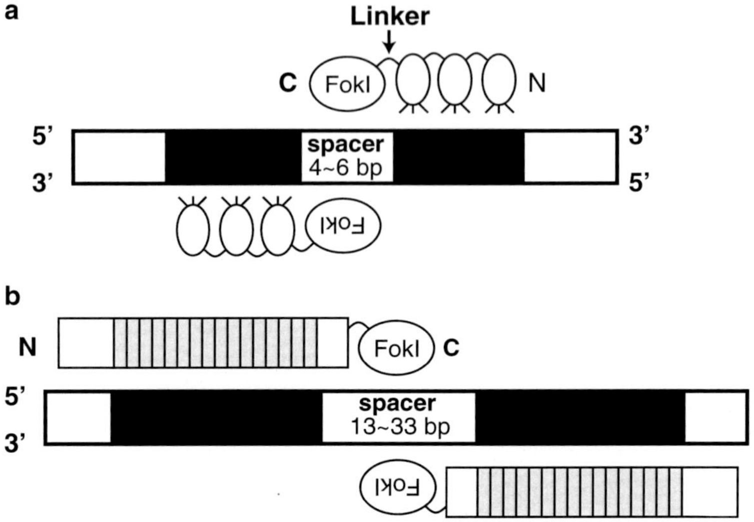

Fig. 1.3 Schematic representation of ZFNs (**a**) and TALENs (**b**). *Black boxes* represent target sequences. Nucleotide sequences between target nucleotides are referred to as spacer sequences

different domains (Li et al. 1992). They fused the FokI nuclease domain with ZFs and created artificial nucleases with novel specificities (Kim et al. 1996). This represented the first demonstration of ZFNs.

FokI needs to be dimerized to cut DNA. A full-length FokI bound to its target site can become dimerized with another unbound FokI in solution and subsequently cleave DNA (Pernstich and Halford 2012). Conversely, perhaps because of the low affinity, the nuclease domains in ZFNs only efficiently dimerize and cut DNA when both ZFNs bind to their target sites separated by the spacer sequence and face one another (Fig. 1.3) (Smith et al. 2000). Thus, ZFNs acquire high specificity from this attribute. However, if not only the target sites for a pair of ZFNs but also palindromes of the target site for one of the ZFNs exist in the genome, both sites could be cleaved by the ZFNs (Fig. 1.4). In the latter case, DSB introduction into unintended sites is called "off-target" cleavage. To avoid this incidence, obligate heterodimeric mutants of FokI nucleases were invented (Fig. 1.4) (Miller et al. 2007; Szczepek et al. 2007; Doyon et al. 2010). The homodimerization of these mutants is significantly inhibited, resulting in a drastic reduction of cytotoxicity (Miller et al. 2007; Szczepek et al. 2007; Doyon et al. 2010). A pair of ZFNs with wild-type (WT) and nuclease-dead mutant FokI nuclease domains cleaves only one strand of DNA (Ramirez et al. 2012; Kim et al. 2012; Wang et al. 2012). Such a single-strand break cannot be repaired through error-prone NHEJ, but is repaired by HDR. Therefore, these "ZFNickases" enable HDR-mediated genome editing almost completely without unwanted off-target effects (Ramirez et al. 2012; Kim et al. 2012; Wang et al. 2012). However, the efficacy of nick-induced HDR is extremely low. ZFs were fused with not only FokI nuclease domains but also other functional domains to generate site-specific activator (Liu et al. 1997), repressor (Liu et al. 1997), methyltransferase (Meister et al. 2010), and recombinase (Sirk et al. 2014).

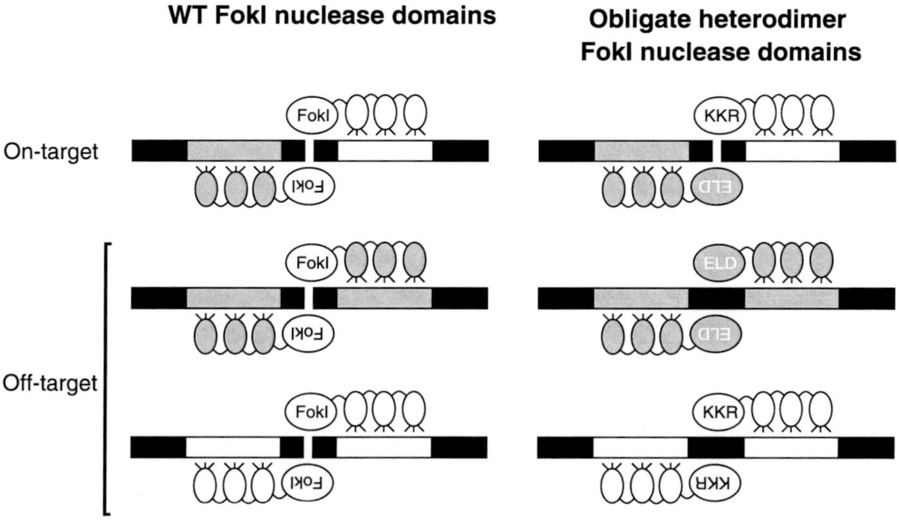

Fig. 1.4 Obligate heterodimeric mutants of FokI nuclease domains reduce off-target effects. Three-finger ZFNs and target DNA are schematically represented. *White and gray colored ZFs* bind to *white and gray target sequences*, respectively. FokI represents the WT FokI nuclease domain. ELD and KKR represent obligate heterodimeric versions of FokI nuclease domains (Doyon et al. 2010). A palindrome of one of the target sites for ZFNs with the WT FokI domain can be cleaved. Such cleavage events at unintended sites are referred to off-target cleavage. On the other hand, ZFNs with an obligate heterodimer of FokI nuclease domains do not cut the palindrome site, thereby significantly reducing the off-target effects

1.3.3 Linkers

It is known that the types of inter-domain linkers between ZF and nuclease domains affect the specificity of ZFNs (Händel et al. 2008). The authors showed that use of a TGAAAR linker restricts the cleavable spacer length to 6 bp (Fig. 1.3). Ambiguity of the cleavable spacer length potentially increases the unintended target sites, leading to higher cytotoxicity.

1.3.4 Engineering Methods of ZFs

In vertebrates, about 2 % of genes encoding proteins contain C_2H_2 ZF domains (Klug 2010). In addition, these domains appear to have the potential to bind to a broad range of nucleotide sequences. Therefore, such domains are potential targets for the design of DNA-binding proteins that bind to user-defined sites. Berg and his colleague successfully showed that ZFs can be engineered to change their specificity (Desjarlais and Berg 1992). Thereafter, several methods for engineering the binding specificity of ZFs were developed as described below. For the engineering of ZFs, mouse ZF268, whose crystal structure had been determined (Kim and Berg 1996), was widely used as the fundamental framework.

1.3.4.1 Sequential Selection of ZFs with Novel Specificity Using a Randomized Library and a Phage Display

As noted earlier, the base recognition mechanisms of ZFs are complicated. Accordingly, selection of ZFs with high specificity and high affinity from a randomized library is an option. Greisman and Pabo constructed a randomized ZF library and selected ZFs with high specificity using a phage display method (Greisman and Pabo 1997). In their randomized library, F1 of ZF268, in which amino acids around the recognition helix (positions −1, +1, +2, +3, +5 and +6) were partially randomized, was joined to the C-terminus of two-finger ZFs comprised of F1 and F2 of ZF268 (Fig. 1.5a) (Greisman and Pabo 1997). Using specific DNA probes as bait, ZFs with high affinity were purified by phage display. Next, F1 at the N-terminus was removed and randomized F2 was added to the C-terminus of the post-screened sublibrary. A second screening was performed using the second library. Finally, a third randomized library was constructed, and ZFs with user-defined binding specificity were screened (Fig. 1.5a). Most of the basic studies of ZF engineering were performed using this method.

1.3.4.2 Bipartite Library and Two-Finger Archive

As mentioned above, the recognition sequences of neighboring ZFs sometimes overlap (Fig. 1.2). Furthermore, inter-finger interactions may also affect the base recognition of ZFs. Therefore, a sequential randomization and selection protocol using phage display may miss the best ZFs with the highest specificity and affinity. To resolve such context dependency, Isaran and colleagues made two new randomized libraries, forming a "bipartite" library (Isalan et al. 2001). In this bipartite library, one had the helical residues of F1 and the N-terminal half of F2 randomized and the other had the helical residues of the C-terminal half of F2 and F3 randomized (Fig. 1.5b) (Isalan et al. 2001). Subsequently, using these libraries and DNA probes containing halves of the target sites, they selected ZFs with high specificity by phage display. After the screening, the N- and C- terminal halves of the sublibraries were joined, and a second screening was performed by phage display. The authors showed that several three-finger ZFs with new binding specificity and high affinity were selected by this method.

To simplify the construction of ZFs with new specificity, the biopharmaceutical company Gendaq (later acquired by Sangamo BioSciences, a leading company in the ZFN technology) constructed a large-number of two-finger ZF archive that recognizes a wide range of DNA sequences (Klug 2010) (Fig. 1.5b). Each two-finger ZF can be combined with a TGEKP linker that does not interfere with its binding properties of each two finger (Moore et al. 2001). Therefore, four- and six-finger ZFs that have higher specificity and affinity can be constructed. Sangamo BioSciences further optimized the binding specificity of ZFs by single amino acid substitutions in the recognition helices (Urnov et al. 2005; Klug 2010). Sigma-Aldrich is currently selling ZFNs that are assembled from the Sangamo database and have been extensively tested for activity.

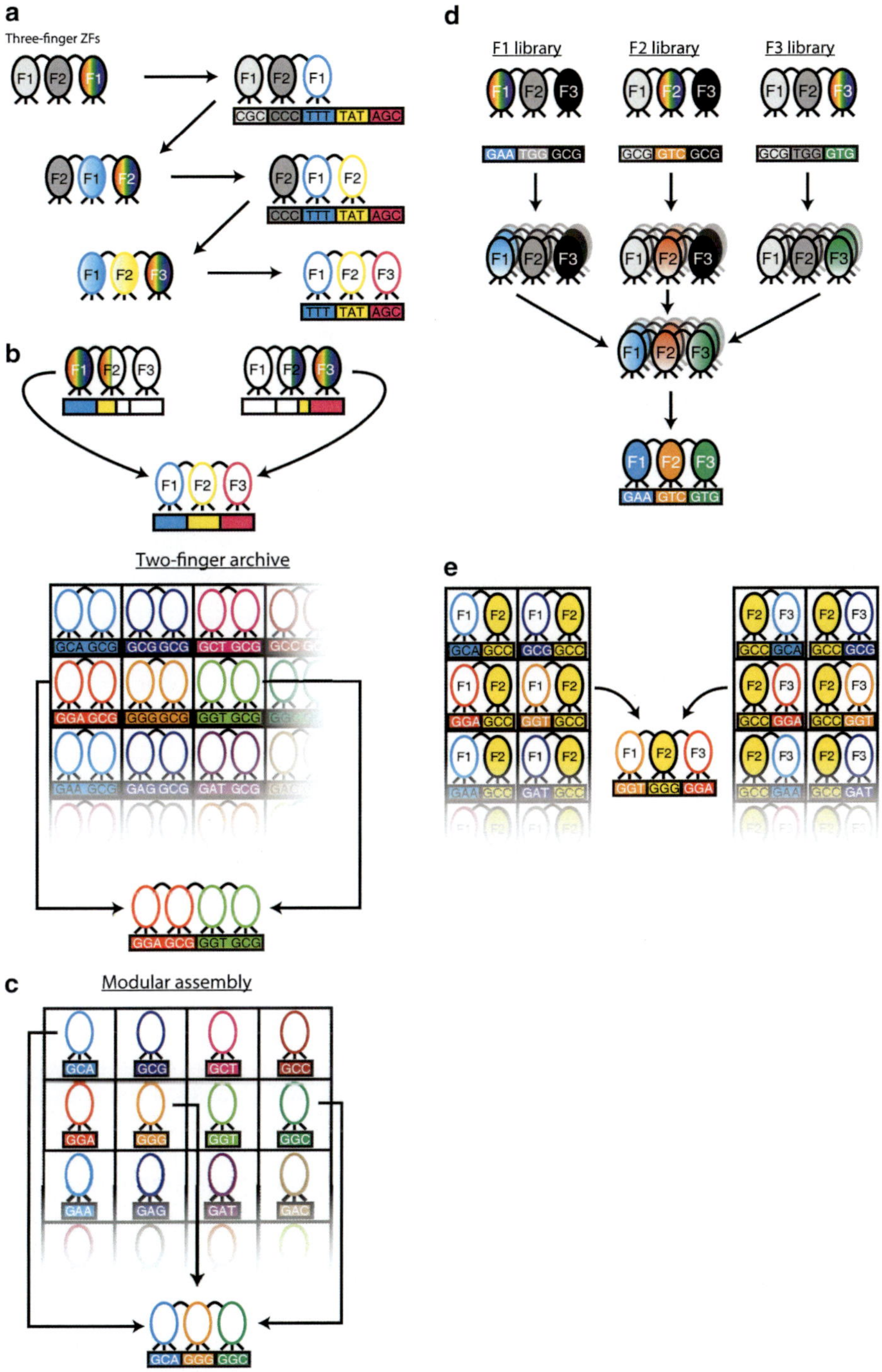

Fig. 1.5 Construction methods for ZFs. Schematic representations of the construction methods are shown. *Color-coded ovals* and *rectangles with solid lines* represent fingers of ZFs and subsites, respectively. ZFs with two-color and multicolor gradients represent selected sublibraries and randomized libraries, respectively. (**a**) Sequential selection of ZFs with novel specificity using a randomized library and phage display. (**b**) Bipartite library and two-finger archive. (**c**) Modular assembly. (**d**) Bacterial-based selection of ZFs from a randomized library. (**e**) Context-dependent assembly (CoDA). See main text for details

1.3.4.3 Modular Assembly

The library construction and screening method with a phage library is time-consuming and labor-intensive. Hence, easier methods are more attractive for most researchers. The "modular assembly" technique is very intuitive and the easiest ZF construction method to date. For this method, three-finger ZFs were assembled using appropriate ZF modules from an archive of one-finger ZFs that recognize various patterns of DNA triplets. ZF domains recognizing GNN (Segal et al. 1999), ANN (Dreier et al. 2001), and CNN (Dreier et al. 2005) were selected using a phage display method. To construct three-finger ZFs recognizing a 9-bp sequence in modular assembly, the user simply joins the appropriate ZF modules together (Fig. 1.5c). Several reports have shown that this method is useful for constructing ZFNs with new binding specificity (Bibikova et al. 2002; Porteus 2006; Beumer et al. 2006; Osakabe et al. 2010). The detailed protocol for the modular assembly and distribution of related plasmids through the nonprofit plasmid vender Addgene appeared to overcome the technical hurdles associated with obtaining user-specific ZFNs for researchers worldwide (Wright et al. 2006). However, after a while, many researchers became aware that the success rate of the creation of functional ZFNs through the modular assembly method is quite low (Ramirez et al. 2008). This appears to arise through context dependency. However, it was reported that the extension of the ZF array length can achieve a better success rate (Bhakta et al. 2013).

1.3.4.4 Bacterial-Based Selection of ZFs from a Randomized Library

In this method, three randomized libraries (F1, F2 and F3 libraries) are initially constructed by randomization of the helical residues in each finger (F1, F2, and F3) of three-finger ZFs fused to a protein that involved in the recruitment of bacterial RNA polymerase (Fig. 1.5d) (Hurt et al. 2003; Meng and Wolfe 2006). Next, *Escherichia coli* containing a reporter plasmid vector, which harbors a ZF recognition sequence upstream of the basal promoter and a selection marker gene, is transformed by each library and plated onto the selection plates. If the ZFs bind to the target site, expression of the selection marker occurs and leads to survival of the bacteria. After collection of plasmid sublibraries from each selection plate (F1, F2, and F3 randomized plates), F1, F2, and F3 from the F1, F2, and F3 sublibraries, respectively, are PCR-amplified, joined together, and inserted into a ZF fusion protein expression vector as the first selection step. The second selection is performed by bacterial-based selection using a selection marker plasmid as the first selection. The surviving colony should have ZFs with binding ability to the target sequence of interest. This method is called the bacterial one-hybrid (Meng and Wolfe 2006; Meng et al. 2008) or bacterial two-hybrid (Hurt et al. 2003) system, depending on the RNA polymerase recruitment strategy. Although three-finger ZFs with new specificity can be efficiently generated by these methods compared with the modular assembly method (Meng et al. 2008; Ochiai et al. 2010; Ochiai et al. 2012), the construction of ZFs by this method is very complicated and takes about 1–2 months. Therefore, this construction method has not been accepted by many researchers.

To simplify the method, Joung and colleagues distributed a preselected archive containing sublibraries from a bacterial-based first selection for several sets of triplet nucleotides, called Oligomerized Pool ENgineering (OPEN) (Maeder et al. 2008). Using this preselected archive, users can generate three-finger ZFs with new specificity by simply joining the ZFs and performing a second selection through bacterial-based selection (Foley et al. 2009; Zou et al. 2009).

1.3.4.5 Context-Dependent Assembly (CoDA)

In the bacterial-based selection method, the first randomization and selection are performed for individual fingers. However, this may miss the best ZFs with new specificity because of the context dependency. To account for this, Joung and colleagues developed context-dependent assembly (CoDA) (Sander et al. 2010). For this method, the authors selected several three-finger ZFs from two randomized libraries. In these libraries, recognition helices of F1 or F3 of three-finger ZFs with F2, which recognizes defined 3-bp subsites, are randomized (Fig. 1.5e). To assemble three-finger ZFs with new specificity, users simply fuse fingers from the two archives (F1-F2 and F2-F3) via the common F2 units (Fig. 1.5e).

1.3.5 Specificity of ZFNs

In general, DNA binding proteins have the potential to bind not to only their most preferred sites (on-target sites) with high affinity, but also to some other sites similar to the on-target sites (off-target sites) with low affinity. In the case of targetable nucleases, these off-target effects may lead to the introduction of unintended mutations and cytotoxicity. Therefore, use of targetable nucleases with high specificity is critical for precise genome editing. Two research groups investigated the off-target effects of ZFNs in human cells and in vitro (Gabriel et al. 2011; Pattanayak et al. 2011). Gabriel et al. used the integrase-defective lentiviral vectors (IDLVs) to infer where the DSBs were introduced in human cultured cells. IDLVs are a linear form in the infected host nucleus and tend to become integrated into the DSB sites within the host genome. The authors infected the IDLVs into a human immortalized myelogenous leukemia line, K562, and investigated where the IDLVs became integrated. Conversely, Gabriel et al. constructed a randomized target library and purified the ZFN proteins. These components were then reacted in vitro. They collected the cleaved fragments and determined what nucleotide sequences were targetable by ZFNs. Both groups used the same *CCR5* ZFNs for off-target analysis and identified the same off-target site in the closely related *CCR2* gene as reported earlier (Perez et al. 2008). Although both groups also found low but significant off-target effects, the obligate heterodimeric version of ZFNs significantly reduced those effects, as expected. Interestingly, the off-target sites detected by in vivo and in vitro analyses at low frequencies were somewhat different. Although the biases from each analytical system seem to be the major causes of the differences, other factors that do not exist in the in vitro system, such as chromatin structure and DNA methylation, may affect the DSB introduction by ZFNs.

The emergence of TALENs, whose construction is much easier than that of ZFNs, has slowed down the expansion of researchers who choose ZFNs for genome editing. However, the four- to six-finger ZFNs created by Sigma-Aldrich and Sangamo BioSciences appear to have high specificity and quite high activity, indicating that ZFNs definitely have great potency.

1.4 TALENs

TALE proteins derived from the bacterial plant pathogen *Xanthomonas* and related genera contain an N-terminal domain, a DNA-binding domain, and a C-terminal domain including an activator domain (Fig. 1.6). The DNA-binding domain

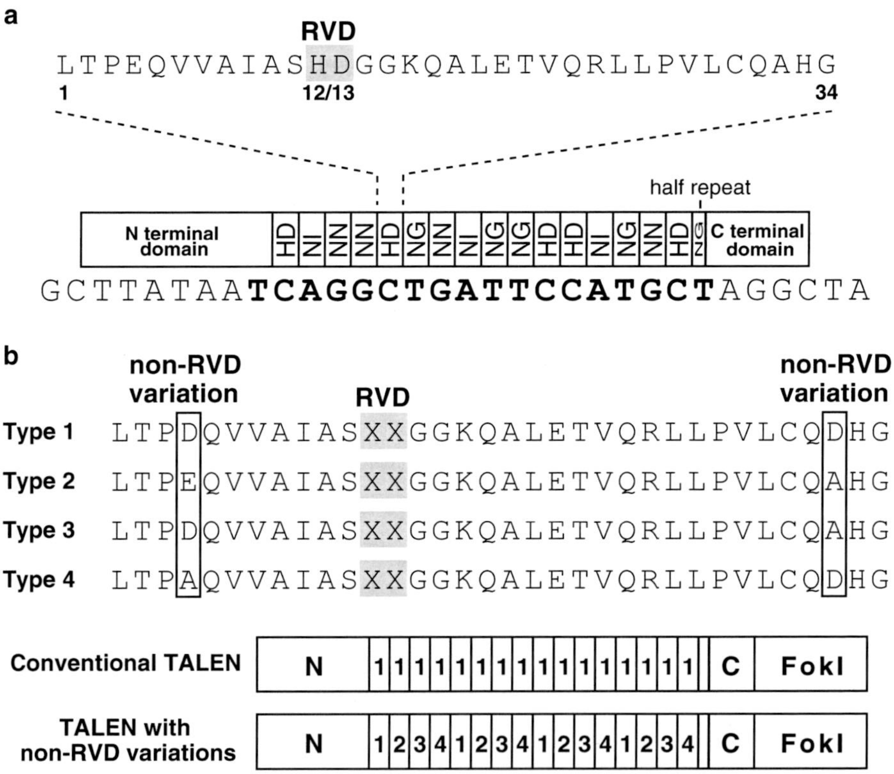

Fig. 1.6 Structure of TALEs. (**a**) Schematic representation of a transcription activator-like effector (TALE) protein and its binding sequence. The amino acid sequence of one of the repeats is shown on the *upper side* of the panel. (**b**) Non-RVD variations at positions 4 and 32 in repeats of TALEs (see main text for details)

composed of multiple nearly identical repeats (7–30), each of which is comprised of approximately 34 amino acids (Fig. 1.6). TALE proteins are injected into host cells through the type III secretion system by the bacteria, bind to the specific DNA sequences, and enhance the expressions of some host genes that enable pathogen spreading (Bogdanove et al. 2010).

In 2009, the DNA recognition mechanisms of TALE proteins were revealed by two research groups (Moscou and Bogdanove 2009; Boch et al. 2009). They performed in silico analyses using a large set of amino acid sequences of DNA-binding domains and their target sequences, and found that the two key amino acids are involved in target recognition. These amino acids are situated in positions 12 and 13 in each 34-amino-acid unit (Fig. 1.6). These positions are highly variable among repeat units and referred to as repeat variable di-residues (RVDs). Each repeat recognizes one nucleotide. The amino acids of RVDs and the recognized nucleotides are well-defined, being NI for A, NG for T, HD for C and NN for G or A (Fig. 1.6). It was reported that the RVD NH can be used to achieve guanine-specific recognition (Cong et al. 2012). The last repeat of the DNA-binding domain contains an RVD, but is only conserved in the N-terminal half of the repeat, the so-called half-repeat. In addition, the N-terminal domain recognizes a thymine adjacent to the nucleotide sequence recognized by the DNA-binding domain. Therefore, the target sequences for TALE proteins start with thymine (Fig. 1.6). Boch et al. showed that TALEs can be redesigned to bind to user-defined sequences by simply joining appropriate repeat units (Boch et al. 2009). Following these reports, several researchers showed the versatilities of these proteins by fusing them with several functional proteins, such as an activator (Zhang et al. 2011), light-stimulated activator (Konermann et al. 2013), repressor (Cong et al. 2012), histone modifier (Mendenhall et al. 2013), DNA demethylase (Maeder et al. 2013), recombinase (Mercer et al. 2012), and of course FokI nuclease (Christian et al. 2010; Li et al. 2011; Miller et al. 2011; Mahfouz et al. 2011). Strictly speaking, several TALE frameworks derived from different origins have been used for various applications. Although the amino acid sequences of the N- and C-terminal domains differed slightly, they are considered to have basically the same properties.

The lengths of the N- and C-terminal domains influence the DSB-introducing activities of TALENs (Miller et al. 2011; Sakuma et al. 2013a). Although several truncation mutants have been reported, TALENs with an N-terminal domain of +136 (or Δ152) and a C-terminal domain of +63 seem to be the most commonly used and highly active (Miller et al. 2011; Hockemeyer et al. 2011; Sander et al. 2011; Tesson et al. 2011; Reyon et al. 2012b). The length of the C-terminal domain also affects the length of the cleavable spacer (Miller et al. 2011; Kim et al. 2013) (Fig. 1.3). The most active +136/+63 architecture can cleave 13–33 bp nucleotide spacers (Christian et al. 2012).

Interestingly, amino acid variations other than the RVD in each repeat unit seem to affect the genome-editing activity of TALENs (Fig. 1.6) (Sander et al. 2011; Reyon et al. 2012b; Sakuma et al. 2013b). Although the TALENs used in most reports have no such non-RVD variations in each repeat unit, some amino acids, particularly those at positions 4 and 32, in natural TALEs have some variation among the repeat units (Fig. 1.6) (Boch and Bonas 2010). Several studies have applied these repeat variations to TALENs and showed a higher success rate and activity compared with conventional TALENs (Miller et al. 2011; Sander et al. 2011; Reyon et al. 2012b). Sakuma et al. compared the activities of TALENs with or without such variations, and found that the TALENs with variations had higher activity than the conventional TALENs (Sakuma et al. 2013b). However, the molecular mechanisms for how the non-RVD variations affect TALEN activity remain an open question.

1.4.1 Construction Methods for TALENs

Compared with ZFs, the recognition code of TALEs is obviously simpler, and does not appear to have any context dependency. In addition, TALEs can be designed to bind to almost any sequence. These properties of TALEs are attractive for many researchers who are interested in genome editing. To simplify the construction of TALEs, many assembly methods have been reported to date. We now introduce some of methods that are frequently used in these research fields.

1.4.1.1 Golden Gate Cloning-Based Assembly

Golden Gate cloning-based assembly uses an archive of TALE repeat units, DNA ligase, and type IIS restriction enzymes, which cut DNA at sites different from the binding site (Cermak et al. 2011) (Fig. 1.7a). Mixing of these enzymes with an appropriate set of plasmids enables the joining 2–10 repeats in a defined order in one reaction. By repeating a similar reaction, it is possible to construct TALEs with 12–31 repeats in 5 days (Cermak et al. 2011). This appears to be the easiest method developed to date, and is used by a large number of researchers around the world (Dahlem et al. 2012; Tong et al. 2012; Bedell et al. 2012; Carlson et al. 2012). Modified versions of the Golden Gate assembly have been reported by various groups (Ding et al. 2013; Kim et al. 2013; Sakuma et al. 2013a, b). Although most of the TALENs constructed by the Golden Gate cloning-based method have repeat units without any non-RVD variations, TALENs with such variations can be constructed by a method based on Golden Gate assembly, called a Platinum Gate assembly (Sakuma et al. 2013b). As mentioned above, these TALENs show higher activity than conventional TALENs. The platinum TALENs enable efficient genome editing in various model systems (Hayashi et al. 2014; Sugi et al. 2014; Ochiai et al. 2014).

1.4.1.2 Golden Gate PCR-Based Assembly

In this method, repeat units are initially amplified using four plasmid encoding repeat units with NN, NG, NI, or HD and appropriate primers with overhangs for Golden Gate assembly (Sanjana et al. 2012) (Fig. 1.7b). The amplified fragments

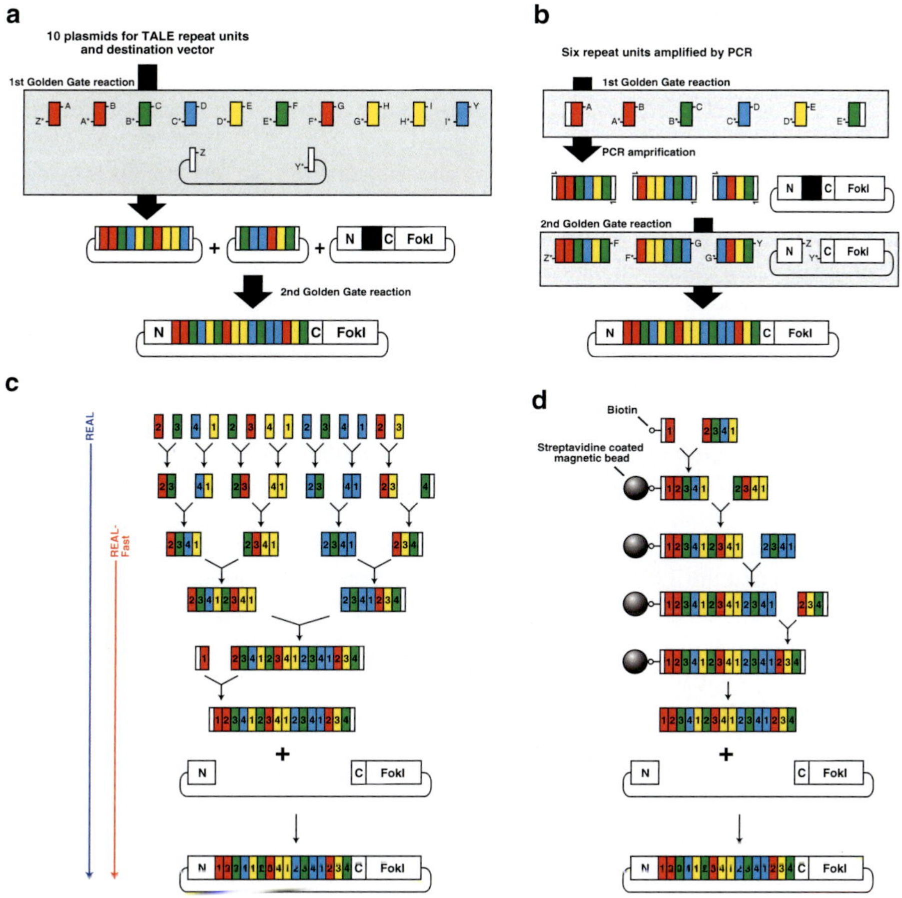

Fig. 1.7 Construction methods for TALEs. Schematic representations of the TALE construction methods are shown. *Colored rectangles* represent repeats of TALEs. *Letters attached to colored rectangles* indicate the types of overhangs generated by type IIS restriction enzymes. *Letters with asterisks* indicate overhangs complementary to ones without asterisks. *Numbers in rectangles* represent types of repeats with non-RVD variations (see main text and Fig. 1.6). (**a**) Golden Gate cloning-based assembly. (**b**) Golden Gate PCR-based assembly. (**c**) Restriction enzyme and ligation (REAL) assembly. Representation of vector backbones has been omitted for clarity. (**d**) Fast ligation-based automatable solid-phase high-throughput (FLASH) system. See main text for details

are purified and used for first-round Golden Gate assembly. The ligated fragments are further amplified by PCR and used for second-round assembly. Compared with the cloning-based assembly, PCR-based assembly requires fewer plasmids and a shorter handling time based on omission of one of the transformation steps. However, purification of PCR products is required for this method, and PCR amplification of nearly identical repeats sometimes leads to mutation of the amplicons.

1.4.1.3 Restriction Enzyme and Ligation (REAL) Assembly

Restriction enzyme and ligation (REAL) assembly is based on type IIS restriction enzymes (Sander et al. 2011) (Fig. 1.7c). In this method, two TALE repeats are initially ligated. Next, each set of two repeats is ligated, and so on (Fig. 1.7c). TALENs with non-RVD variations can be constructed by choosing appropriate repeat units provided by the authors (Reyon et al. 2012a). REAL is the simplest method for assembling TALENs, but requires much more time than the other methods because of the multiple cloning steps. An improved version of the method, REAL-Fast, in which 2 to 4-unit preassembled modules are used instead of single units, was recently reported (Reyon et al. 2012a). Although it requires a shorter time to assemble TALEs than the REAL method, the pre-assembled archive is quite large (more than 350 plasmids).

1.4.1.4 Fast Ligation-Based Automatable Solid-Phase High-Throughput (FLASH) System

In this method, almost the same repeat units are used as in REAL-Fast. The major difference is that the TALE repeats are assembled on solid-phase magnetic beads in the FLASH system (Reyon et al. 2012b) (Fig. 1.7d). Initially, a biotin-labeled N-terminal unit and a 4-unit pre-assembled module are ligated, and anchored to streptavidin-coated magnetic beads. After purification, serial digestions and ligations are performed to assemble TALEs with non-RVD variations as in REAL-Fast. With automation of the process, the authors claimed the assembly of 96 different TALE repeats in less than 1 day (Reyon et al. 2012b).

1.4.2 Specificity of TALENs

TALENs seem to have higher genome editing activity and less toxicity than ZFNs (Chen et al. 2013). This might arise through the high specificity—in other words, the low off-target effects—of TALENs. Recently, the specificity of TALENs was analyzed in vitro, similar to ZFNs (Guilinger et al. 2014). The authors showed that each TALE repeat independently recognizes its corresponding nucleotide base, and despite a tolerance of mismatches of longer TALENs in vitro, they showed sufficient specificity in human cells. Moreover, they found that excessive binding energy of TALENs toward DNA reduces the specificity.

1.5 Applications of ZFNs and TALENs

In this section, we will briefly introduce some examples of targetable nuclease-mediated genome editing. This technology is dependent on the introduction of DSBs into user-intended sites. Therefore, outcome achieved by one type of targetable nuclease can also be accomplished with other types of nucleases. The details of each example will be provided in other chapters of this book.

As mentioned earlier, NHEJ-mediated gene disruption has been used in various model systems (Carroll 2014). This enables the exploration of deeper aspects of the gene functions. In addition, Sangamo BioSciences is trying to cure acquired immunodeficiency syndrome (AIDS) induced by human immunodeficiency virus (HIV) through NHEJ-mediated disruption of *CCR5*, which is involved in HIV infection in T-cells, using ZFNs. They are evaluating the outcomes in clinical trials. In a recent report (Tebas et al. 2014), the authors collected $CD4^+$ T-cells from patients and introduced *CCR5* ZFNs into these cells. The cells, of which 11–28 % were genetically modified by the ZFNs, were then autologously infused. They found that the blood levels of HIV DNA decreased in most patients and concluded that *CCR5-modified* $CD4^+$ T-cell infusions are safe within the limits of the study (Tebas et al. 2014).

Simultaneous cutting of two genomic sites can induce large deletions (Lee et al. 2010), inversions (Gupta et al. 2013), and translocations (Piganeau et al. 2013) through NHEJ repair. In addition, simultaneous cutting of the target genomic site and a plasmid vector facilitates vector integration into the site through NHEJ (Maresca et al. 2012).

HDR-mediated genome editing enables more complicated and precise genome modifications than NHEJ-mediated genome editing. To introduce a selection marker or GFP into a specific genomic site, a plasmid vector containing exogenous element flanked by about 1-kb homology arms has been used for conventional HDR-mediated genome editing (Moehle et al. 2007). However, DSBs are mainly repaired through NHEJ. Thus, in most of the cases, drug selection is required to efficiently establish cell clones in which HDR-mediated genome editing has occurred (Shukla et al. 2009; Zou et al. 2009; Soldner et al. 2011). Consequently, compared with cultured cells, relatively few reports have shown successful HDR-mediated genome editing in multicellular organisms, in which efficient methods for selection of recombinant cells are hardly available (Beumer et al. 2008; Meyer et al. 2010; Cui et al. 2010; Ochiai et al. 2012; Zu et al. 2013).

HDR-mediated genome editing can also introduce single-nucleotide substitutions (Urnov et al. 2005; Soldner et al. 2011; Yusa et al. 2011; Ochiai et al. 2014; Miyaoka et al. 2014). This not only enables the establishment of disease models (Yusa et al. 2011; Ochiai et al. 2014), but also has potential for the performance of gene therapy.

Single-stranded oligodeoxynucleotides (ssODNs) as well as double-strand DNA can serve as a repair template for HDR (Soldner et al. 2011; Chen et al. 2011; Meyer et al. 2012; Bedell et al. 2012). Compared with conventional targeting vectors, ssODNs are not principally integrated into the host genome. Co-introduction of targetable nucleases and ssODNs enables efficient introduction of small nucleotide changes (single nucleotide substitution, or introduction of short DNA elements, such as a *loxP* site and restriction enzyme sites).

1.6 Summary

In this chapter, we have mainly focused on the properties and construction methods of ZFNs and TALENs. In general, construction of ZFNs with high specificity are not simple compared with TALENs and CRISPR/Cas (see Chap. 2). In addition, several reports have suggested that TALENs have higher specificity than three-finger ZFNs (Dahlem et al. 2012; Chen et al. 2013). Therefore, many researchers chose TALENs, not ZFNs, for genome editing until the emergence of the CRISPR/Cas system, whose preparation for genome editing is the easiest among the targetable nucleases. However, TALENs seem to have much higher specificity than CRISPR/Cas (Fu et al. 2013). Therefore, TALENs are still important tools for genome editing. Some of the four- to six-finger ZFNs, most of which are constructed by Sangamo BioSciences or Sigma-Aldrich, have excellent specificity and genome editing activities (Shukla et al. 2009; Hockemeyer et al. 2009; Mashimo et al. 2010; Soldner et al. 2011). As noted, Sangamo BioSciences is using ZFNs for HIV therapy (Tebas et al. 2014), suggesting that ZFNs have high potential for genome editing.

References

Aitman TJ, Critser JK, Cuppen E et al (2008) Progress and prospects in rat genetics: a community view. Nat Genet 40:516–522

Bedell VM, Wang Y, Campbell JM et al (2012) In vivo genome editing using a high-efficiency TALEN system. Nature 491:114–118

Beumer K, Bhattacharyya G, Bibikova M et al (2006) Efficient gene targeting in *Drosophila* with zinc-finger nucleases. Genetics 172:2391–2403

Beumer KJ, Trautman JK, Bozas A et al (2008) Efficient gene targeting in *Drosophila* by direct embryo injection with zinc-finger nucleases. Proc Natl Acad Sci U S A 105:19821–19826

Bhakta MS, Henry IM, Ousterout DG et al (2013) Highly active zinc-finger nucleases by extended modular assembly. Genome Res 23:530–538

Bibikova M, Golic M, Golic KG, Carroll D (2002) Targeted chromosomal cleavage and mutagenesis in *Drosophila* using zinc-finger nucleases. Genetics 161:1169–1175

Boch J, Bonas U (2010) Xanthomonas AvrBs3 family-type III effectors: discovery and function. Annu Rev Phytopathol 48:419–436

Boch J, Scholze H, Schornack S et al (2009) Breaking the code of DNA binding specificity of TAL-type III effectors. Science 326:1509–1512

Bogdanove AJ, Schornack S, Lahaye T (2010) TAL effectors: finding plant genes for disease and defense. Curr Opin Plant Biol 13:394–401

Carlson DF, Tan W, Lillico SG et al (2012) Efficient TALEN-mediated gene knockout in livestock. Proc Natl Acad Sci U S A 109:17382–17387

Carroll D (2014) Genome engineering with targetable nucleases. Annu Rev Biochem 83:409–439

Cermak T, Doyle EL, Christian M et al (2011) Efficient design and assembly of custom TALEN and other TAL effector-based constructs for DNA targeting. Nucleic Acids Res 39:e82–e82

Chen F, Pruett-Miller SM, Huang Y et al (2011) High-frequency genome editing using ssDNA oligonucleotides with zinc-finger nucleases. Nat Methods 8:753–755

Chen S, Oikonomou G, Chiu CN et al (2013) A large-scale in vivo analysis reveals that TALENs are significantly more mutagenic than ZFNs generated using context-dependent assembly. Nucleic Acids Res 41:2769–2778

Christian M, Cermak T, Doyle EL et al (2010) Targeting DNA double-strand breaks with TAL effector nucleases. Genetics 186:757–761

Christian ML, Demorest ZL, Starker CG et al (2012) Targeting G with TAL effectors: a comparison of activities of TALENs constructed with NN and NK repeat variable di-residues. PLoS One 7:e45383

Cong L, Zhou R, Kuo Y-C et al (2012) Comprehensive interrogation of natural TALE DNA-binding modules and transcriptional repressor domains. Nat Commun 3:968

Cong L, Ran FA, Cox D et al (2013) Multiplex genome engineering using CRISPR/Cas systems. Science 339:819–823

Cui X, Ji D, Fisher DA et al (2010) Targeted integration in rat and mouse embryos with zinc-finger nucleases. Nat Biotechnol 29:64–67

Dahlem TJ, Hoshijima K, Jurynec MJ et al (2012) Simple methods for generating and detecting locus-specific mutations induced with TALENs in the zebrafish genome. PLoS Genet 8:e1002861

Desjarlais JR, Berg JM (1992) Redesigning the DNA-binding specificity of a zinc finger protein: a data base-guided approach. Proteins 12:101–104

Ding Q, Lee Y-K, Schaefer EAK et al (2013) A TALEN genome-editing system for generating human stem cell-based disease models. Cell Stem Cell 12:238–251

Doyon Y, Vo TD, Mendel MC et al (2010) Enhancing zinc-finger-nuclease activity with improved obligate heterodimeric architectures. Nat Methods 8:74–79

Dreier B, Beerli RR, Segal DJ et al (2001) Development of zinc finger domains for recognition of the 5"-ANN-3" family of DNA sequences and their use in the construction of artificial transcription factors. J Biol Chem 276:29466–29478

Dreier B, Fuller RP, Segal DJ et al (2005) Development of zinc finger domains for recognition of the 5'-CNN-3' family DNA sequences and their use in the construction of artificial transcription factors. J Biol Chem 280:35588–35597

Emery DW (2011) The use of chromatin insulators to improve the expression and safety of integrating gene transfer vectors. Hum Gene Ther 22:761–774

Foley JE, Yeh J-RJ, Maeder ML et al (2009) Rapid mutation of endogenous zebrafish genes using zinc finger nucleases made by Oligomerized Pool ENgineering (OPEN). PLoS One 4:e4348

Fu Y, Foden JA, Khayter C et al (2013) High-frequency off-target mutagenesis induced by CRISPR-Cas nucleases in human cells. Nat Biotechnol 31:822–826

Gabriel R, Lombardo A, Arens A et al (2011) An unbiased genome-wide analysis of zinc-finger nuclease specificity. Nat Biotechnol 29:816–823

Greisman HA, Pabo CO (1997) A general strategy for selecting high-affinity zinc finger proteins for diverse DNA target sites. Science 275:657–661

Guilinger JP, Pattanayak V, Reyon D et al (2014) Broad specificity profiling of TALENs results in engineered nucleases with improved DNA-cleavage specificity. Nat Methods 11:429–435

Gupta A, Hall VL, Kok FO et al (2013) Targeted chromosomal deletions and inversions in zebrafish. Genome Res 23:1008–1017

Hayashi T, Sakamoto K, Sakuma T et al (2014) Transcription activator-like effector nucleases efficiently disrupt the target gene in Iberian ribbed newts (Pleurodeles waltl), an experimental model animal for regeneration. Dev Growth Differ 56:115–121

Händel E-M, Alwin S, Cathomen T (2008) Expanding or restricting the target site repertoire of zinc-finger nucleases: the inter-domain linker as a major determinant of target site selectivity. Mol Ther 17:104–111

Hockemeyer D, Soldner F, Beard C et al (2009) Efficient targeting of expressed and silent genes in human ESCs and iPSCs using zinc-finger nucleases. Nat Biotechnol 27:851–857

Hockemeyer D, Wang H, Kiani S et al (2011) Genetic engineering of human pluripotent cells using TALE nucleases. Nat Biotechnol 29:731–734

Hurt JA, Thibodeau SA, Hirsh AS et al (2003) Highly specific zinc finger proteins obtained by directed domain shuffling and cell-based selection. Proc Natl Acad Sci U S A 100:12271–12276

Isalan M, Klug A, Choo Y (2001) A rapid, generally applicable method to engineer zinc fingers illustrated by targeting the HIV-1 promoter. Nat Biotechnol 19:656–660

Ivics Z, Li MA, Mátés L et al (2009) Transposon-mediated genome manipulation in vertebrates. Nat Methods 6:415–422
Joung JK, Sander JD (2013) TALENs: a widely applicable technology for targeted genome editing. Nat Rev Mol Cell Biol 14:49–55
Kim CA, Berg JM (1996) A 2.2 A resolution crystal structure of a designed zinc finger protein bound to DNA. Nat Struct Biol 3:940–945
Kim YG, Cha J, Chandrasegaran S (1996) Hybrid restriction enzymes: zinc finger fusions to Fok I cleavage domain. Proc Natl Acad Sci U S A 93:1156–1160
Kim E, Kim S, Kim DH et al (2012) Precision genome engineering with programmable DNA-nicking enzymes. Genome Res 22:1327–1333
Kim Y, Kweon J, Kim A et al (2013) A library of TAL effector nucleases spanning the human genome. Nat Biotechnol 31:251–258
Klug A (2010) The discovery of zinc fingers and their applications in gene regulation and genome manipulation. Annu Rev Biochem 79:213–231
Konermann S, Brigham MD, Trevino AE et al (2013) Optical control of mammalian endogenous transcription and epigenetic states. Nature 500:472–476
Lee HJ, Kim E, Kim J-S (2010) Targeted chromosomal deletions in human cells using zinc finger nucleases. Genome Res 20:81–89
Li L, Wu LP, Chandrasegaran S (1992) Functional domains in Fok I restriction endonuclease. Proc Natl Acad Sci U S A 89:4275–4279
Li T, Huang S, Jiang WZ et al (2011) TAL nucleases (TALNs): hybrid proteins composed of TAL effectors and FokI DNA-cleavage domain. Nucleic Acids Res 39:359–372
Liu Q, Segal DJ, Ghiara JB, Barbas CF (1997) Design of polydactyl zinc-finger proteins for unique addressing within complex genomes. Proc Natl Acad Sci U S A 94:5525–5530
Maeder ML, Thibodeau-Beganny S, Osiak A et al (2008) Rapid "open-source" engineering of customized zinc-finger nucleases for highly efficient gene modification. Mol Cell 31: 294–301
Maeder ML, Angstman JF, Richardson ME et al (2013) Targeted DNA demethylation and activation of endogenous genes using programmable TALE-TET1 fusion proteins. Nat Biotechnol 31:1137–1142
Mahfouz MM, Li L, Shamimuzzaman M et al (2011) De novo-engineered transcription activator-like effector (TALE) hybrid nuclease with novel DNA binding specificity creates double-strand breaks. Proc Natl Acad Sci U S A 108:2623–2628
Mali P, Esvelt KM, Church GM (2013a) Cas9 as a versatile tool for engineering biology. Nat Methods 10:957–963
Mali P, Yang L, Esvelt KM et al (2013b) RNA-guided human genome engineering via Cas9. Science 339:823–826
Maresca M, Lin VG, Guo N, Yang Y (2012) Obligate ligation-gated recombination (ObLiGaRe): custom designed nucleases mediated targeted integration through non-homologous end joining. Genome Res 23:539–546
Mashimo T, Takizawa A, Voigt B et al (2010) Generation of knockout rats with X-linked severe combined immunodeficiency (X-SCID) using zinc-finger nucleases. PLoS One 5:e8870
Megason SG, Fraser SE (2007) Imaging in systems biology. Cell 130:784–795
Meister GE, Chandrasegaran S, Ostermeier M (2010) Heterodimeric DNA methyltransferases as a platform for creating designer zinc finger methyltransferases for targeted DNA methylation in cells. Nucleic Acids Res 38:1749–1759
Mendenhall EM, Williamson K, Reyon D et al (2013) Locus-specific editing of histone modifications at endogenous enhancers. Nat Biotechnol 31:1133–1136
Meng X, Wolfe SA (2006) Identifying DNA sequences recognized by a transcription factor using a bacterial one-hybrid system. Nat Protoc 1:30–45
Meng X, Noyes MB, Zhu LJ et al (2008) Targeted gene inactivation in zebrafish using engineered zinc-finger nucleases. Nat Biotechnol 26:695–701
Mercer AC, Gaj T, Fuller RP, Barbas CF (2012) Chimeric TALE recombinases with programmable DNA sequence specificity. Nucleic Acids Res 40:11163–11172

Meyer M, de Angelis MH, Wurst W, Kühn R (2010) Gene targeting by homologous recombination in mouse zygotes mediated by zinc-finger nucleases. Proc Natl Acad Sci U S A 107:15022–15026

Meyer M, Ortiz O, Hrabé de Angelis M et al (2012) Modeling disease mutations by gene targeting in one-cell mouse embryos. Proc Natl Acad Sci U S A 109:9354–9359

Miller JC, Holmes MC, Wang J et al (2007) An improved zinc-finger nuclease architecture for highly specific genome editing. Nat Biotechnol 25:778–785

Miller JC, Tan S, Qiao G et al (2011) A TALE nuclease architecture for efficient genome editing. Nat Biotechnol 29:143–148

Miyaoka Y, Chan AH, Judge LM et al (2014) Isolation of single-base genome-edited human iPs cells without antibiotic selection. Nat Methods 11:291–293

Moehle EA, Moehle EA, Rock JM et al (2007) Targeted gene addition into a specified location in the human genome using designed zinc finger nucleases. Proc Natl Acad Sci U S A 104: 3055–3060

Moore M, Klug A, Choo Y (2001) Improved DNA binding specificity from polyzinc finger peptides by using strings of two-finger units. Proc Natl Acad Sci U S A 98:1437–1441

Moscou MJ, Bogdanove AJ (2009) A simple cipher governs DNA recognition by TAL effectors. Science 326:1501–1501

Ochiai H, Fujita K, Suzuki K-I et al (2010) Targeted mutagenesis in the sea urchin embryo using zinc-finger nucleases. Genes Cells 15:875–885

Ochiai H, Sakamoto N, Fujita K et al (2012) Zinc-finger nuclease-mediated targeted insertion of reporter genes for quantitative imaging of gene expression in sea urchin embryos. Proc Natl Acad Sci U S A 109:10915–10920

Ochiai H, Miyamoto T, Kanai A et al (2014) TALEN-mediated single-base-pair editing identification of an intergenic mutation upstream of BUB1B as causative of PCS (MVA) syndrome. Proc Natl Acad Sci U S A 111:1461–1466

Osakabe K, Osakabe Y, Toki S (2010) Site-directed mutagenesis in *Arabidopsis* using custom-designed zinc finger nucleases. Proc Natl Acad Sci U S A 107:12034–12039

Pattanayak V, Ramirez CL, Joung JK, Liu DR (2011) Revealing off-target cleavage specificities of zinc-finger nucleases by in vitro selection. Nat Methods 8:765–770

Perez EE, Wang J, Miller JC et al (2008) Establishment of HIV-1 resistance in CD4+ T cells by genome editing using zinc-finger nucleases. Nat Biotechnol 26:808–816

Pernstich C, Halford SE (2012) Illuminating the reaction pathway of the FokI restriction endonuclease by fluorescence resonance energy transfer. Nucleic Acids Res 40:1203–1213

Piganeau M, Ghezraoui H, De Cian A et al (2013) Cancer translocations in human cells induced by zinc finger and TALE nucleases. Genome Res 23:1182–1193

Porteus MH (2006) Mammalian gene targeting with designed zinc finger nucleases. Mol Ther 13:438–446

Ramirez CL, Foley JE, Wright DA et al (2008) Unexpected failure rates for modular assembly of engineered zinc fingers. Nat Methods 5:374–375

Ramirez CL, Certo MT, Mussolino C et al (2012) Engineered zinc finger nickases induce homology-directed repair with reduced mutagenic effects. Nucleic Acids Res 40:5560–5568

Reyon D, Khayter C, Regan MR et al. (2012a) Engineering designer transcription activator-like effector nucleases (TALENs) by REAL or REAL-Fast assembly. Curr Protoc Mol Biol Chapter 12:Unit 12.15. doi:10.1002/0471142727.mb1215s100

Reyon D, Tsai SQ, Khayter C et al (2012b) FLASH assembly of TALENs for high-throughput genome editing. Nat Biotechnol 30:460–465

Sakuma T, Hosoi S, Woltjen K et al (2013a) Efficient TALEN construction and evaluation methods for human cell and animal applications. Genes Cells 18:315–326

Sakuma T, Ochiai H, Kaneko T et al (2013b) Repeating pattern of non-RVD variations in DNA-binding modules enhances TALEN activity. Sci Rep 3:3379

Sander JD, Dahlborg EJ, Goodwin MJ et al (2010) Selection-free zinc-finger-nuclease engineering by context-dependent assembly (CoDA). Nat Methods 8:67–69

Sander JD, Cade L, Khayter C et al (2011) Targeted gene disruption in somatic zebrafish cells using engineered TALENs. Nat Biotechnol 29:697–698

Sanjana NE, Cong L, Zhou Y et al (2012) A transcription activator-like effector toolbox for genome engineering. Nat Protoc 7:171–192

Segal DJ, Dreier B, Beerli RR, Barbas CF (1999) Toward controlling gene expression at will: selection and design of zinc finger domains recognizing each of the 5"-GNN-3" DNA target sequences. Proc Natl Acad Sci U S A 96:2758–2763

Shukla VK, Doyon Y, Miller JC et al (2009) Precise genome modification in the crop species Zea mays using zinc-finger nucleases. Nature 459:437–441

Sirk SJ, Gaj T, Jonsson A et al (2014) Expanding the zinc-finger recombinase repertoire: directed evolution and mutational analysis of serine recombinase specificity determinants. Nucleic Acids Res 42:4755–4766

Smith J, Bibikova M, Whitby FG et al (2000) Requirements for double-strand cleavage by chimeric restriction enzymes with zinc finger DNA-recognition domains. Nucleic Acids Res 28: 3361–3369

Soldner F, Laganière J, Cheng AW et al (2011) Generation of isogenic pluripotent stem cells differing exclusively at Two early onset Parkinson point mutations. Cell 146:318–331

Subramanya S, Kim S-S, Manjunath N, Shankar P (2010) RNA interference-based therapeutics for human immunodeficiency virus HIV-1 treatment: synthetic siRNA or vector-based shRNA? Expert Opin Biol Ther 10:201–213

Sugi T, Sakuma T, Ohtani Y, Yamamoto T (2014) Versatile strategy for isolating transcription activator-like effector nuclease-mediated knockout mutants in *Caenorhabditis elegans*. Dev Growth Differ 56:78–85

Szczepek M, Brondani V, Büchel J et al (2007) Structure-based redesign of the dimerization interface reduces the toxicity of zinc-finger nucleases. Nat Biotechnol 25:786–793

Tebas P, Stein D, Tang WW et al (2014) Gene editing of CCR5in autologous CD4 T cells of persons infected with HIV. N Engl J Med 370:901–910

Tesson L, Usal C, Ménoret S et al (2011) Knockout rats generated by embryo microinjection of TALENs. Nat Biotechnol 29:695–696

Tong C, Huang G, Ashton C et al (2012) Rapid and cost-effective gene targeting in rat embryonic stem cells by TALENs. J Genet Genomics 39:275–280

Urnov FD, Miller JC, Lee Y-L et al (2005) Highly efficient endogenous human gene correction using designed zinc-finger nucleases. Nature 435:646–651

Urnov FD, Rebar EJ, Holmes MC et al (2010) Genome editing with engineered zinc finger nucleases. Nat Rev Genet 11:636–646

Wang J, Friedman G, Doyon Y et al (2012) Targeted gene addition to a predetermined site in the human genome using a ZFN-based nicking enzyme. Genome Res 22:1316–1326

Wolfe SA, Nekludova L, Pabo CO (2000) DNA recognition by Cys2His2 zinc finger proteins. Annu Rev Biophys Biomol Struct 29:183–212

Wright DA, Thibodeau-Beganny S, Sander JD et al (2006) Standardized reagents and protocols for engineering zinc finger nucleases by modular assembly. Nat Protoc 1:1637–1652

Yates L, McMurray F, Zhang Y et al (2009) ENU mutagenesis as a tool for understanding lung development and disease. Biochem Soc Trans 37:838–842

Yusa K, Rashid ST, Strick-Marchand H et al (2011) Targeted gene correction of α1-antitrypsin deficiency in induced pluripotent stem cells. Nature 478:391–394

Zhang F, Cong L, Lodato S et al (2011) Efficient construction of sequence-specific TAL effectors for modulating mammalian transcription. Nat Biotechnol 29:149–153

Zou J, Maeder ML, Mali P et al (2009) Gene targeting of a disease-related gene in human induced pluripotent stem and embryonic stem cells. Cell Stem Cell 5:97–110

Zu Y, Tong X, Wang Z et al (2013) TALEN-mediated precise genome modification by homologous recombination in zebrafish. Nat Methods 10:329–331

Chapter 2
CRISPR/Cas9: The Leading Edge of Genome Editing Technology

Tetsushi Sakuma and Takashi Yamamoto

Abstract An RNA-guided endonuclease (RGEN), known as CRISPR/Cas9, has been dramatically changing the field of genome engineering. Because CRISPR/Cas9 is much easier to introduce than ZFNs or TALENs because of its simple construction of customized vectors targeting particular genomic loci, this epoch-making technology has rapidly become a standard tool for targeted gene modification within a time span of just a few years. In this chapter, we explain how the technology has arisen, how it has become established, improved, and applied, and how it will evolve in the future. CRISPR/Cas9-mediated genome editing strategies are likely to continue to accelerate studies on functional genomics for years to come. Moreover, nuclease-inactivated Cas9 (dCas9) with various functional domains will develop the technology to its fullest potential, in addition to ZF- and TALE-based platforms. CRISPR/Cas9 will change the face not only of genetic engineering, but also of a variety of research areas in life science studies.

Keywords Cas9 nuclease • CRISPR/Cas9 • Genome editing • Genome engineering • gRNA • Guide RNA • RGEN • RNA-guided endonuclease

2.1 Introduction

Engineered endonuclease-mediated genome editing using ZFNs and TALENs is conceptually similar to restriction endonuclease-mediated DNA manipulation in vitro. Given that recent advances in the engineering of programmable nucleases have almost completely abolished any limitations of the target sequences, especially for TALENs, it appears as though the use of restriction endonucleases has become unrestricted.

On the other hand, engineered endonuclease-mediated genome editing strategies always require the construction of customized nucleases corresponding to the

T. Sakuma (✉) • T. Yamamoto
Department of Mathematical and Life Sciences, Graduate School of Science, Hiroshima University, Higashi-Hiroshima, Hiroshima 739-8526, Japan
e-mail: tetsushi-sakuma@hiroshima-u.ac.jp

T. Yamamoto (ed.), *Targeted Genome Editing Using Site-Specific Nucleases*,
DOI 10.1007/978-4-431-55227-7_2

intended genomic sequence. As an analogy, the principle is similar to the performance of immunostaining using primary antibodies directly conjugated with alkaline phosphatase. As we know, two components, primary antibodies without any conjugations and secondary antibodies recognizing the primary antibodies conjugated with alkaline phosphatase, can facilitate immunochemical manipulation, because the phosphatase-conjugated secondary antibody can be used for any primary antibodies recognizing a variety of antigens. The same phenomenon is also expected in genome editing technology, and prokaryotic immunity has represented a "diamond in the rough" to realize such two-component gene targeting.

2.2 The CRISPR/Cas System in Prokaryotic Adaptive Immunity

The clustered regularly interspaced short palindromic repeats (CRISPR) locus is found in the genomes of some bacteria and archaea (Ishino et al. 1987; Mojica et al. 2000; Jansen et al. 2002). It contains tandem repeats and spacers, in which the repeats comprise the same sequence and the spacers comprise different sequences derived from exotic DNA (Mojica et al. 2005; Pourcel et al. 2005). The CRISPR locus functions with CRISPR-associated (Cas) proteins as an adaptive immune system against invading foreign DNA (CRISPR/Cas system; Fig. 2.1) (Wiedenheft et al. 2012; Westra et al. 2014). In the system, the invading DNA is incorporated into the spacer region in the CRISPR locus and transcribed as a long pre-crRNA (CRISPR RNA) containing multiple repeats and spacers. Subsequently, in the type II CRISPR/Cas system, pre-crRNA is processed to crRNA harboring a single spacer sequence complementary to the foreign DNA with another short RNA molecule, trans-crRNA (tracrRNA), transcribed from a different locus. The resulting crRNA–tracrRNA heteroduplex works as a guidance molecule to target exogenous DNA with the identical sequence to the crRNA, and induce a DNA double-strand break (DSB) at the specific locus in association with Cas protein(s) (Bhaya et al. 2011; Reeks et al. 2013; Barrangou and Marraffini 2014).

2.3 Application of CRISPR/Cas9 in Genome Editing

When applying the CRISPR/Cas system in genome editing, only two components are needed, namely a chimeric guide RNA (gRNA) mimicking the crRNA–tracrRNA complex, and a Cas9 protein with nuclease activity (Fig. 2.2) (Jinek et al. 2012; Cong et al. 2013; Mali et al. 2013a). Although the targeting specificity is mainly dependent on the gRNA sequence, Cas9 also requires a few particular bases, known as a protospacer adjacent motif (PAM) (Bolotin et al. 2005). The PAM sequences vary among species. For example, *Streptococcus pyogenes* Cas9 (SpCas9) requires 5′-NGG-3′ (Jinek et al. 2012), *Streptococcus thermophilus* Cas9 (StCas9) requires

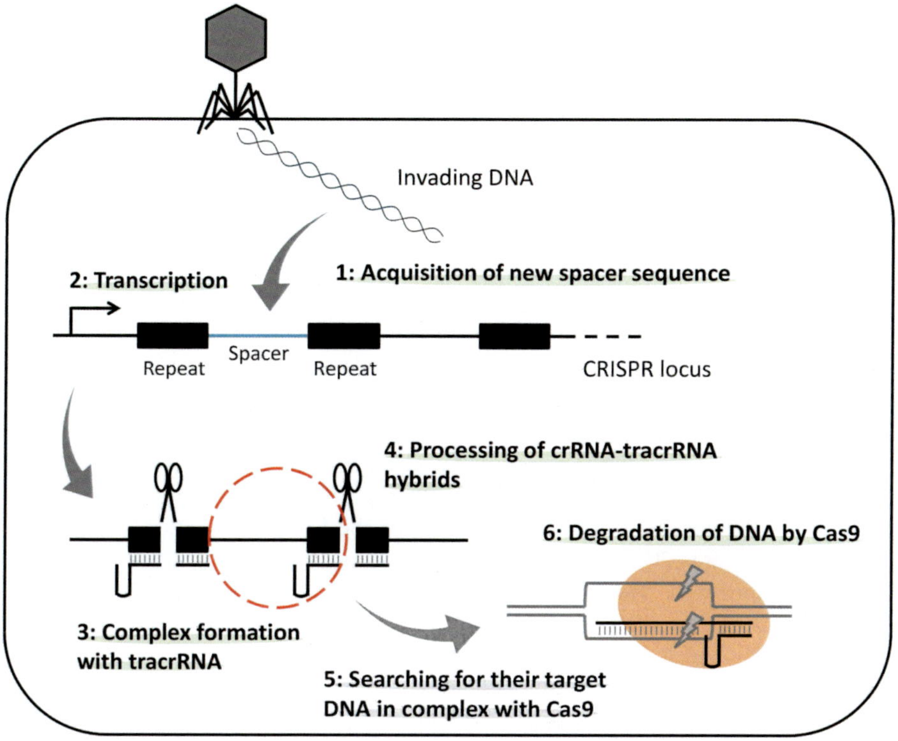

Fig. 2.1 Natural mechanism of the type II CRISPR/Cas adaptive immune system. Foreign nucleotides such as viral DNA or plasmids are incorporated into the CRISPR locus on the host genome. Following new spacer acquisition, pre-crRNA is transcribed and hybridized with tracrRNA. After processing, the crRNA–tracrRNA complex is recruited to the target DNA sequence along with Cas9 protein, and degradation occurs through the nuclease activity of Cas9

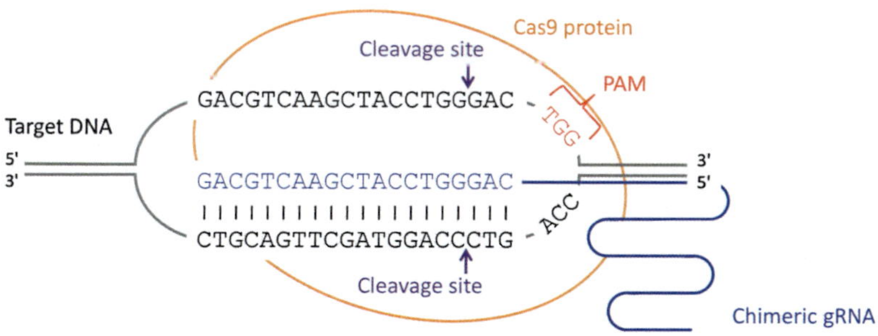

Fig. 2.2 Schematic representation for target DNA recognition and cleavage by a gRNA-Cas9 complex. SpCas9 initially searches for the PAM sequence (5′-NGG-3′) on the target DNA. Subsequently, base-pairing between the target DNA and gRNA gradually occurs from the PAM side. After 20-bp hybridization, the target DNA is cleaved by Cas9 nuclease, resulting in a blunt end at the 3-bp upstream of the PAM site

5′-NNAGAAW-3′ (Cong et al. 2013; Esvelt et al. 2013), and *Neisseria meningitidis* Cas9 (NmCas9) requires 5′-NNNNGATT-3′ (Esvelt et al. 2013; How et al. 2013; Walsh and Hochedlinger 2013). Currently, SpCas9 is the most widely used for genetic engineering (Hsu et al. 2014; Wilkinson and Wiedenheft 2014). The SpCas9-gRNA complex is known to initially seek out the PAM sequence in the genome, and subsequently unwind the double-stranded DNA and form DNA-RNA base-pairing in a directional manner (Sternberg et al. 2014). When introducing a DSB, two nuclease domains, HNH and RuvC, independently induce a nick at the Watson and Crick strands, resulting in a linear DSB between the bases at 3- and 4-bp upstream of the PAM sequence (Jinek et al. 2012; Nishimasu et al. 2014).

The gRNA structure is another important factor for CRISPR/Cas9-based genome editing. Although crRNA and tracrRNA can be separately transcribed like the naturally-occurring CRISPR/Cas system, a chimeric gRNA structure is rather simple and often leads to high activity (Hsu et al. 2013). A chimeric gRNA consists of a crRNA-derived region at the 5′ end and a tracrRNA-derived region at the 3′ end, and various modifications have been adopted in both regions by several groups (reviewed in Sander and Joung 2014). Basically, the DNA-recognition sequence in the crRNA region is 20-bp long, but the addition or truncation of a few bases can reportedly improve the specificity (Cho et al. 2014; Fu et al. 2014). The 3′ end of the crRNA region and the 5′ end of the tracrRNA region are generally linked with four nucleotides (5′-GAAA-3′) to form a major stem loop, known as a tetraloop (Kim and Kim 2014). Stem extensions have also been reported (Chen et al. 2013; Hsu et al. 2013; Jinek et al. 2013). The tracrRNA region has additional minor loops on the 3′ side, and these sequences are known to be important for high gRNA expression (Hsu et al. 2013). In addition, A-U flips in the poly-A or poly-T regions have been adopted in some studies (Chen et al. 2013; Jinek et al. 2013).

2.4 Targeting Specificity of CRISPR/Cas9

As described above, the approximately 20-bp gRNA sequence and 3-bp PAM sequence of SpCas9 define the targeting specificity of CRISPR/Cas9. However, the stringency of base recognition is not equivalent among these sequences. Regarding the PAM sequence, SpCas9 has the ability to bind to 5′-NGA-3′ (Zhang et al. 2014) and 5′-NAG-3′ (Hsu et al. 2013; Jiang et al. 2013) sites as well as 5′-NGG-3′. Regarding the gRNA targeting sequence, the specificity decreases with increasing distance from the PAM site. The sequence extending up to 12 bp adjacent to the PAM site is called the seed sequence, and has relatively high targeting specificity (Jinek et al. 2012; Cong et al. 2013).

In some types of cultured cells, especially immortalized cell lines such as U2OS, HEK293T, and K562, highly frequent off-target mutations have been observed by many groups (Cradick et al. 2013; Fu et al. 2013; Hsu et al. 2013; Pattanayak et al. 2013; Cho et al. 2014; Lin et al. 2014). In vitro assays have also shown off-target binding with high frequencies (Pattanayak et al. 2013). However, in normal cells such as mouse embryonic stem (mES) cells and organisms such as mice and rats, the levels of induced off-target mutations do not seem to be as high (Wang et al. 2013; Mashiko et al. 2014; Yoshimi et al. 2014). Furthermore, whole-genome sequencing has recently been conducted by several groups to analyze the off-target mutations in genome-edited cells, resulting in findings that individual cell clones had low frequencies of unintended mutations among human stem cells treated with CRISPR/Cas9, as well as TALENs (Smith et al. 2014; Suzuki et al. 2014; Veres et al. 2014). These reports suggest that the mutation frequencies at off-target sites vary among species and cell types because of potential differences in the DSB repair machineries.

Interestingly, a genome-wide survey of SpCas9-gRNA-binding sites using chromatin immunoprecipitation followed by sequencing (ChIP-seq) revealed that only seven nucleotides, including 5′-GG-3′ in the PAM, could be identified as a consensus sequence (Wu et al. 2014). Furthermore, although thousands of off-target binding sites were determined, only one of the analyzed potential off-target sites carried significant mutations in mES cells (1/295; 0.34 %). Similar results were reported by two other groups (Duan et al. 2014; Kuscu et al. 2014). On the other hand, 70 % of SpCas9-gRNA-binding sites were found to be associated with genes (Wu et al. 2014). This result suggests that changes in transcriptional regulation can be triggered by CRISPR/Cas9 for unintended genes, because various studies have shown that binding of catalytically inactive Cas9 to a coding region or regulatory region causes transcriptional inhibition (CRISPRi) (Gilbert et al. 2013; Larson et al. 2013; Qi et al. 2013; Zhao et al. 2014). Although further studies are needed to clarify this issue, we need to recognize the possibility of such potential side effects without any mutations when using CRISPR/Cas9.

2.5 Double-Nicking and Dimeric FokI-dCas9 Strategies for Highly-Specific Genome Editing

Based on the strong concern about off-target mutations, several advanced strategies for highly specific CRISPR/Cas9-mediated genome editing have been developed (Fig. 2.3). The main problem for CRISPR/Cas9 specificity is its monomeric architecture, unlike the case for the dimeric ZFNs and TALENs. A conventional CRISPR/Cas9 genome editing system contains a single gRNA and a Cas9 nuclease. Since the Cas9 nuclease has cleavage activity for both DNA strands, the induced DSB site is determined by the single gRNA.

Previous research on engineered endonucleases has provided some clues to solve the problem of specificity. The TALE::TevI architecture, known as compact TALEN (cTALEN), can induce a nick when used as a monomer, but can also induce a DSB when used as a pair (Beurdeley et al. 2013). This paired nicking can only cleave DNA when the space between two nicks is within a range of defined lengths (9–18 bp). Similarly, the double nicking induced by CRISPR/Cas9 was reported to introduce a DSB (Mali et al. 2013b; Ran et al. 2013; Cho et al. 2014). The nuclease activity of Cas9 can be converted to nickase activity when a D10A or H840A mutation is incorporated (Jinek et al. 2012). Theoretically, these nickases cannot induce a DSB unless two adjacent nicks on both strands are introduced. Therefore, when using

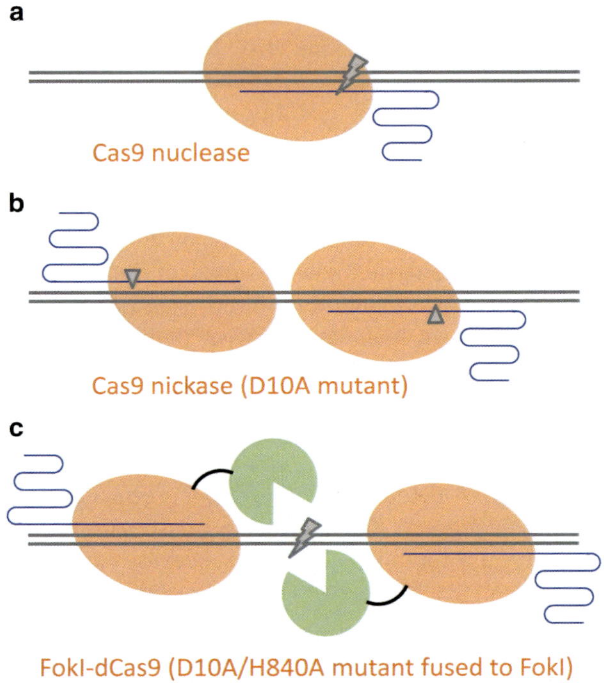

Fig. 2.3 Three different DNA-cleaving strategies using the CRISPR/Cas9 system. (**a**) Original CRISPR/Cas9 system mediated by wild-type Cas9 nuclease and a single gRNA. (**b**) Double-nicking strategy mediated by Cas9 nickase harboring the D10A mutation and two gRNAs. (**c**) RNA-guided FokI nuclease system mediated by catalytically inactive Cas9 harboring D10A and H840A mutations (dCas9) fused to the nuclease domain of FokI and two gRNAs

Cas9 nickase, the target site needs to be recognized by two distinct gRNAs, meaning that targeted mutagenesis can be performed in a highly specific manner. Moreover, this double-nicking strategy reportedly works not only in cultured cells, but also in embryos of animals such as mice (Fujii et al. 2014; Shen et al. 2014).

Another attempt to improve the specificity involves the creation of a fusion protein between catalytically inactive Cas9 (dCas9) and the nuclease domain of FokI (RNA-guided FokI nuclease; RFN). dCas9 has both D10A and H840A mutations and no DNA-cleaving activity. Tsai et al. (2014) showed that FokI-dCas9 can be used as a dimeric nuclease similar to ZFNs and TALENs. FokI-dCas9 can introduce a DSB when the spacer length is in the range of 13–18 bp. In fact, paired nicking is not truly a dimeric strategy, because Cas9 nickase is catalytically active and the nicks sometimes induce mutations (Tsai et al. 2014). On the other hand, FokI-dCas9 acts as a proper dimeric nuclease. The applicability of RFNs in various organisms other than cultured cells needs to be investigated in future studies.

2.6 Web-Based Software for Designing gRNA Targets and Predicting Off-Target Candidates

In principle, the sequence limitation for targeting with CRISPR/Cas9 is only a PAM site. In practice, other actual limitations are as follows: 1) sequences harboring poly-T should be avoided as gRNA target sequences, because poly-T can work as a transcriptional terminator; and 2) the number of potential off-target sites should be minimized.

Currently, various tools for designing CRISPR/Cas9 target sites and predicting potential off-target sites are available on the web. CRISPR Design Tool (http://crispr.mit.edu/), developed by the Feng Zhang laboratory at MIT (Ran et al. 2013), is supposedly the most widely used resource for designing and assessing gRNA target sequences (Ni et al. 2014; Yen et al. 2014; Yoshimi et al. 2014). CRISPR Design Tool is used not only for design, but also for searching for off-target sites in the genomes of certain species, including humans, rats, mice, zebrafish, flies, and nematodes. ZiFiT Targeter (http://zifit.partners.org/ZiFiT/) (Sander et al. 2007, 2010; Hwang et al. 2013; Fu et al. 2014) available on the website of the Zinc Finger Consortium is also commonly used for designing gRNAs (Blitz et al. 2013; Nakayama et al. 2013; Yu et al. 2014). Other web resources include E-CRISP (http://www.e-crisp.org/E-CRISP/) (Heigwer et al. 2014), CRISPRdirect (http://crispr.dbcls.jp/), CRISPR Optimal Target Finder (http://tools.flycrispr.molbio.wisc.edu/targetFinder/) (Gratz et al. 2014), CasOT (http://eendb.zfgenetics.org/casot/) (Xiao et al. 2014), Cas-OFFinder (http://www.rgenome.net/cas-offinder/) (Bae et al. 2014), CHOPCHOP (https://chopchop.rc.fas.harvard.edu/) (Montague et al. 2014), sgRNAcas9 (http://www.biootools.com/col.jsp?id=103) (Xie et al. 2014), and CRISPy (http://staff.biosustain.dtu.dk/laeb/crispy/) (Ronda et al. 2014).

CRISPR Genome Analyzer, CRISPR-GA (http://crispr-ga.net/) (Guell et al. 2014), developed by the George Church laboratory, is a different type of web tool. Using CRISPR-GA, we can obtain analytical data by uploading forward and reverse reads of Miseq sequences from the amplicons of genetically modified cells or organisms. The percentages of error-prone non-homologous end-joining are calculated, and the sizes and locations of deletions and insertions can be visualized in automatically created figures.

2.7 Construction of CRISPR/Cas9 Vectors

Plasmids for constructing custom gRNA- and Cas9-expressing vectors are available from Addgene (https://www.addgene.org/) and several other commercial companies including Life Technologies, OriGene, and System BioSciences. The construction procedure only involves insertion of annealed oligonucleotides into the vectors, which is much simpler than the procedures for ZFNs or TALENs (Fig. 2.4) (Cong et al. 2013; Ran et al. 2013). The gRNA and Cas9 can be expressed using either separate vectors or a single combined vector. pX330, a single vector expressing both gRNA and Cas9 nuclease with human U6 and chicken beta-actin hybrid (CBh) promoters, respectively, was originally developed by the Feng Zhang laboratory (Cong et al. 2013) and has been very widely used for cell and animal genome editing (Mashiko et al. 2013, 2014; Ran et al. 2013; Matsunaga and Yamashita 2014; Mizuno et al. 2014; Park et al. 2014; Yin et al. 2014).

Meanwhile, Sakuma et al. (2014) developed an all-in-one CRISPR/Cas9 vector system for multiplex genome engineering by modifying the pX330 plasmid. In their

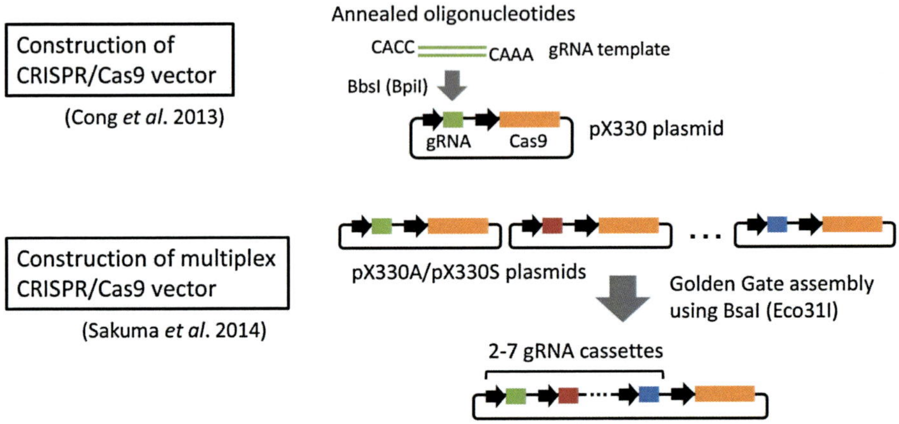

Fig. 2.4 Construction methods for CRISPR/Cas9 vectors for single (*upper panel*) and multiple (*lower panel*) gene targeting. pX330, originally developed in the Feng Zhang laboratory, is probably the most commonly used CRISPR/Cas9 vector. A template DNA sequence for the gRNA should be prepared as annealed oligonucleotides and inserted into the BbsI-digested pX330 vector. For multiplex genome engineering, an all-in-one vector system containing multiple gRNA cassettes and a Cas9 cassette can be used. The system involves the BsaI-mediated Golden Gate assembly method for the concatemerization of gRNA cassettes

system, up to seven gRNA expression cassettes are tandemly ligated into a single vector along with a Cas9 nuclease/nickase cassette using the Golden Gate assembly method (Fig. 2.4), which is often used for modular assembly of DNA-binding repeats of TALE (Cermak et al. 2011; Kim et al. 2013; Sakuma et al. 2013a, b). The all-in-one vector constructed with this system has been proven to be applicable for simultaneous gene targeting of up to seven and three genomic loci with standard nuclease and paired nickase strategies, respectively. The materials for the construction are expected be distributed as the "Multiplex CRISPR/Cas9 Assembly System Kit" by Addgene.

2.8 Methods for Introducing CRISPR/Cas9 into Cells and Organisms

To achieve CRISPR/Cas9-mediated genome engineering, various methodologies have been devised and conducted for delivery of the two components, gRNA and Cas9. For cultured cells and animal embryos, the two components can be introduced by DNA/RNA/protein transfection or microinjection. Multiplex genome engineering is also applicable when multiple plasmids, plasmid and DNA fragments, single all-in-one plasmids, or RNA/protein are introduced (Fig. 2.5) (Jao et al. 2013; Li et al. 2013b; Wang et al. 2013; Guo et al. 2014; Ma et al. 2014; Sakuma et al. 2014). Purified Cas9 protein and gRNAs can form ribonucleoproteins (RNPs) in vitro,

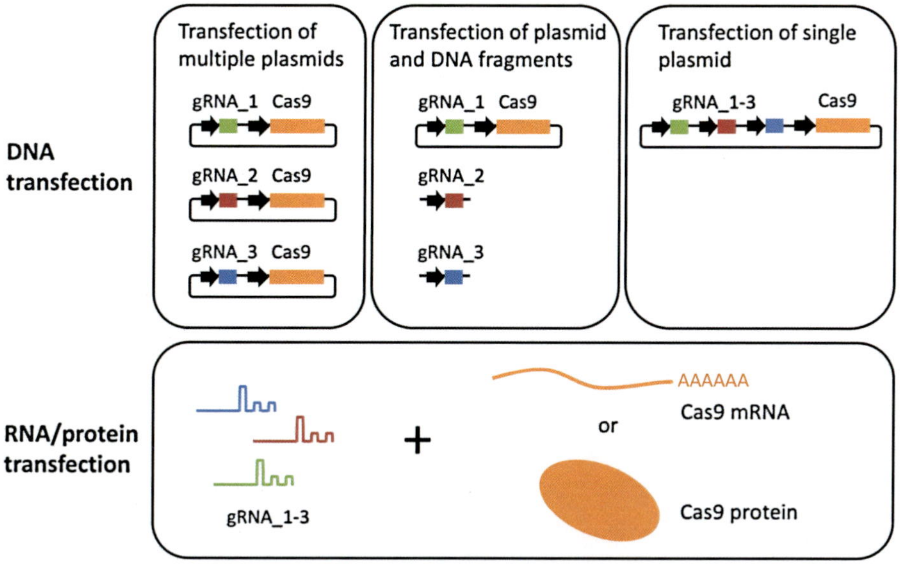

Fig. 2.5 Examples of transfection strategies for CRISPR/Cas9-mediated multiplex genome engineering. Multiple plasmids, plasmid and DNA fragments, or single plasmids can be applied for the DNA transfection (*upper panels*). Alternatively, Csy4-mediated cleavage of long transcripts can produce multiple gRNAs (Nissim et al. 2014; Tsai et al. 2014). Several gRNAs transcribed in vitro and Cas9 mRNA or protein can be used for DNA-free transfection

which can be incorporated into cells by electroporation (Kim et al. 2014b). If a cell-penetrating peptide is added for gRNAs and conjugated with Cas9, RNPs can be delivered into cells by simply adding them into the medium (Ramakrishna et al. 2014), similar to the case for TALENs with cell-penetrating peptides (Ru et al. 2013; Liu et al. 2014). Lentiviral delivery into cells and animals has also been reported (Malina et al. 2013; Heckl et al. 2014). Importantly, a lentiviral CRISPR/Cas9 library has enabled forward genetics screening in cultured cells (Koike-Yusa et al. 2014; Shalem et al. 2014; Wang et al. 2014; Zhou et al. 2014). For plant applications, protoplast transformation or *Agrobacteria* infection has generally been used for the delivery (Feng et al. 2013; Li et al. 2013a; Nekrasov et al. 2013; Shan et al. 2013). The current situations for CRISPR/Cas9-mediated genome editing in various cells and organisms are described in Part II of this book.

2.9 Expanded Applications of CRISPR/Cas9 in Life Science Studies

Similar to ZF- and TALE-based technologies, fusion proteins of dCas9 with various functional domains can act as a variety of site-specific DNA-binding effector proteins (Fig. 2.6) (Mali et al. 2013c; Hsu et al. 2014; Sander and Joung 2014).

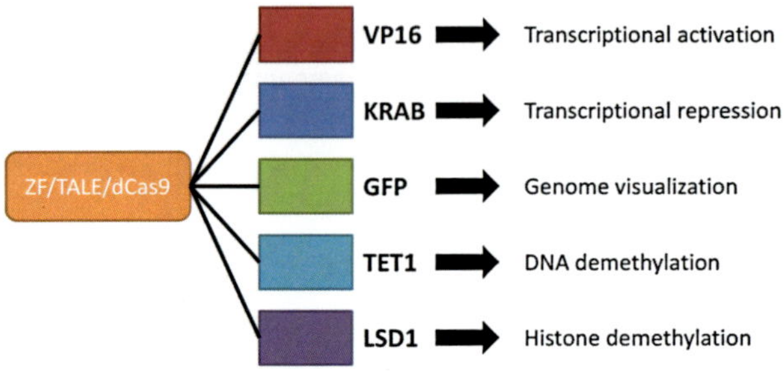

Fig. 2.6 Various applications of ZF/TALE/CRISPR technologies. Genome editing techniques can expand beyond site-specific nucleases. For example, VP16 and KRAB fusions result in transcriptional activation and repression of specific genes, respectively, GFP fusion results in visualization of specific genomic loci, and TET1 and LSD1 fusions result in site-specific epigenetic modifications

For transcriptional activation, the herpes simplex virus-derived activator domain, VP16, or its concatemers, such as VP48, VP64, and VP120, are fused with dCas9 (Cheng et al. 2013; Maeder et al. 2013a; Hu et al. 2014). Regarding transcriptional repression, although dCas9 itself can inhibit transcription (Gilbert et al. 2013; Larson et al. 2013; Qi et al. 2013; Zhao et al. 2014), dCas9 fused with a repressor domain such as KRAB can repress gene expression more efficiently (Gilbert et al. 2013; Kearns et al. 2014). dCas9-GFP, developed by Chen et al. (2103), enables dynamic imaging of genomic loci in cultured cells. Site-specific epigenome editing is also thought to be applicable using a dCas9-fusion strategy (Rusk 2014), but only a few examples have currently been reported using TALE-based strategies (Konermann et al. 2013; Maeder et al. 2013b; Mendenhall et al. 2013) and there are no reports for CRISPR technology. Nuclease-independent genetic engineering enzymes have also been adopted in ZF/TALE-fusion architectures. ZF/TALE-recombinases and ZF/TALE-transposases have been reported in the following papers: Gordley et al. (2007), Gersbach et al. (2011), Mercer et al. (2012), and Gaj et al. 2013 for recombinases; Li et al. (2013c) and Owens et al. (2013) for transposases. CRISPR applications for these purposes are expected to be developed in the near future.

It is particularly worth noting that CRISPR/Cas9-based transcriptional control methodologies open up a huge new field of synthetic biology, as well as TALE-based transcriptional modulation techniques (Farzadfard et al. 2013; Kiani et al. 2014; Moore et al. 2014). Among others, Nissim et al. (2014) constructed particularly sophisticated gene networks using CRISPR transcriptional control tools in combination with various RNA-modifying systems such as RNA-triple-helix structures, introns, microRNAs, and ribozymes. In the meantime however, further expansion and deepening of the technologies are required for this challenging field.

In addition, there are some other applications that differ from the standard genome engineering approaches. Engineered DNA-binding molecule-mediated chromatin immunoprecipitation, enChIP, is one of the unique methods utilizing CRISPR/Cas9 (Fujita and Fujii 2013). Using enChIP, specific genomic regions can be efficiently purified and their associated proteins can be identified by mass spectrometry. The same group further showed that similar experiments can be performed with TALEs instead of CRISPR/Cas9 (Fujita et al. 2013). Kim et al. (2014a) applied the CRISPR/Cas9 system to the genotyping of polymorphisms in vitro. Restriction fragment length polymorphism (RFLP) analysis is often used for genotyping of genome-edited alleles (Suzuki et al. 2013; Nakagawa et al. 2014; Sakane et al. 2014). However, conventional RFLP analysis can only be applied when there is a recognition sequence for a restriction enzyme around the DSB site. On the other hand, the CRISPR/Cas9-mediated RFLP method using in vitro-synthesized gRNA and Cas9 protein enables the genotyping of any sequence, as long as there is a PAM sequence around the target site.

Conclusions

Owing to its simplicity, convenience, and flexibility, CRISPR/Cas9 technology is currently evolving with astonishing rapidity (Pennisi 2013). Along with novel mechanistic insights, various improvements and upgrades of the system are continuously being reported, together with a rich variety of applications that are too numerous to mention. This innovative technology is clearly upsetting conventional wisdom in every research field in life science studies. Researchers are encouraged to enjoy the benefits of this novel technique and drive the growth of their science.

References

Bae S, Park J, Kim JS (2014) Cas-OFFinder: a fast and versatile algorithm that searches for potential off-target sites of Cas9 RNA-guided endonucleases. Bioinformatics 30:1473–1475

Barrangou R, Marraffini LA (2014) CRISPR-Cas systems: prokaryotes upgrade to adaptive immunity. Mol Cell 54:234–244

Beurdeley M, Bietz F, Li J, Thomas S, Stoddard T, Juillerat A, Zhang F, Voytas DF, Duchateau P, Silva GH (2013) Compact designer TALENs for efficient genome engineering. Nat Commun 4:1762

Bhaya D, Davison M, Barrangou R (2011) CRISPR-Cas systems in bacteria and archaea: versatile small RNAs for adaptive defense and regulation. Annu Rev Genet 45:273–297

Blitz IL, Biesinger J, Xie X, Cho KW (2013) Biallelic genome modification in F_0 *Xenopus tropicalis* embryos using the CRISPR/Cas system. Genesis 51:827–834

Bolotin A, Quinquis B, Sorokin A, Ehrlich SD (2005) Clustered regularly interspaced short palindrome repeats (CRISPRs) have spacers of extrachromosomal origin. Microbiology 151: 2551–2561

Cermak T, Doyle EL, Christian M, Wang L, Zhang Y, Schmidt C, Baller JA, Somia NV, Bogdanove AJ, Voytas DF (2011) Efficient design and assembly of custom TALEN and other TAL effector-based constructs for DNA targeting. Nucleic Acids Res 39:e82

Chen B, Gilbert LA, Cimini BA, Schnitzbauer J, Zhang W, Li GW, Park J, Blackburn EH, Weissman JS, Qi LS, Huang B (2013) Dynamic imaging of genomic loci in living human cells by an optimized CRISPR/Cas system. Cell 155(7):1479–1491

Cheng AW, Wang H, Yang H, Shi L, Katz Y, Theunissen TW, Rangarajan S, Shivalila CS, Dadon DB, Jaenisch R (2013) Multiplexed activation of endogenous genes by CRISPR-on, an RNA-guided transcriptional activator system. Cell Res 23:1163–1171

Cho SW, Kim S, Kim Y, Kweon J, Kim HS, Bae S, Kim JS (2014) Analysis of off-target effects of CRISPR/Cas-derived RNA-guided endonucleases and nickases. Genome Res 24(1):132–141

Cong L, Ran FA, Cox D, Lin S, Barretto R, Habib N, Hsu PD, Wu X, Jiang W, Marraffini LA, Zhang F (2013) Multiplex genome engineering using CRISPR/Cas systems. Science 339:819–823

Cradick TJ, Fine EJ, Antico CJ, Bao G (2013) CRISPR/Cas9 systems targeting β-globin and CCR5 genes have substantial off-target activity. Nucleic Acids Res 41:9584–9592

Duan J, Lu G, Xie Z, Lou M, Luo J, Guo L, Zhang Y (2014) Genome-wide identification of CRISPR/Cas9 off-targets in human genome. Cell Res. doi:10.1038/cr.2014.87

Esvelt KM, Mali P, Braff JL, Moosburner M, Yaung SJ, Church GM (2013) Orthogonal Cas9 proteins for RNA-guided gene regulation and editing. Nat Methods 10:1116–1121

Farzadfard F, Perli SD, Lu TK (2013) Tunable and multifunctional eukaryotic transcription factors based on CRISPR/Cas. ACS Synth Biol 2:604–613

Feng Z, Zhang B, Ding W, Liu X, Yang DL, Wei P, Cao F, Zhu S, Zhang F, Mao Y, Zhu JK (2013) Efficient genome editing in plants using a CRISPR/Cas system. Cell Res 23:1229–1232

Fu Y, Foden JA, Khayter C, Maeder ML, Reyon D, Joung JK, Sander JD (2013) High-frequency off-target mutagenesis induced by CRISPR-Cas nucleases in human cells. Nat Biotechnol 31:822–826

Fu Y, Sander JD, Reyon D, Cascio VM, Joung JK (2014) Improving CRISPR-Cas nuclease specificity using truncated guide RNAs. Nat Biotechnol 32:279–284

Fujii W, Onuma A, Sugiura K, Naito K (2014) Efficient generation of genome-modified mice via offset-nicking by CRISPR/Cas system. Biochem Biophys Res Commun 445:791–794

Fujita T, Fujii H (2013) Efficient isolation of specific genomic regions and identification of associated proteins by engineered DNA-binding molecule-mediated chromatin immunoprecipitation (enChIP) using CRISPR. Biochem Biophys Res Commun 439:132–136

Fujita T, Asano Y, Ohtsuka J, Takada Y, Saito K, Ohki R, Fujii H (2013) Identification of telomere-associated molecules by engineered DNA-binding molecule-mediated chromatin immunoprecipitation (enChIP). Sci Rep 3:3171

Gaj T, Mercer AC, Sirk SJ, Smith HL, Barbas CF 3rd (2013) A comprehensive approach to zinc-finger recombinase customization enables genomic targeting in human cells. Nucleic Acids Res 41:3937–3946

Gersbach CA, Gaj T, Gordley RM, Mercer AC, Barbas CF 3rd (2011) Targeted plasmid integration into the human genome by an engineered zinc-finger recombinase. Nucleic Acids Res 39:7868–7878

Gilbert LA, Larson MH, Morsut L, Liu Z, Brar GA, Torres SE, Stern-Ginossar N, Brandman O, Whitehead EH, Doudna JA, Lim WA, Weissman JS, Qi LS (2013) CRISPR-mediated modular RNA-guided regulation of transcription in eukaryotes. Cell 154:442–451

Gordley RM, Smith JD, Graslund T, Barbas CF 3rd (2007) Evolution of programmable zinc finger-recombinases with activity in human cells. J Mol Biol 367:802–813

Gratz SJ, Ukken FP, Rubinstein CD, Thiede G, Donohue LK, Cummings AM, O'Connor-Giles KM (2014) Highly specific and efficient CRISPR/Cas9-catalyzed homology-directed repair in *Drosophila*. Genetics 196:961–971

Guell M, Yang L, Church G (2014) Genome editing assessment using CRISPR genome analyzer (CRISPR-GA). Bioinformatics. doi:10.1093/bioinformatics/btu427

Guo X, Zhang T, Hu Z, Zhang Y, Shi Z, Wang Q, Cui Y, Wang F, Zhao H, Chen Y (2014) Efficient RNA/Cas9-mediated genome editing in *Xenopus tropicalis*. Development 141:707–714

Heckl D, Kowalczyk MS, Yudovich D, Belizaire R, Puram RV, McConkey ME, Thielke A, Aster JC, Regev A, Ebert BL (2014) Generation of mouse models of myeloid malignancy with combinatorial genetic lesions using CRISPR-Cas9 genome editing. Nat Biotechnol. doi:10.1038/nbt.2951

Heigwer F, Kerr G, Boutros M (2014) E-CRISP: fast CRISPR target site identification. Nat Methods 11:122–123

Hou Z, Zhang Y, Propson NE, Howden SE, Chu LF, Sontheimer EJ, Thomson JA (2013) Efficient genome engineering in human pluripotent stem cells using Cas9 from *Neisseria meningitidis*. Proc Natl Acad Sci U S A 110:15644–15649

Hsu PD, Scott DA, Weinstein JA, Ran FA, Konermann S, Agarwala V, Li Y, Fine EJ, Wu X, Shalem O, Cradick TJ, Marraffini LA, Bao G, Zhang F (2013) DNA targeting specificity of RNA-guided Cas9 nucleases. Nat Biotechnol 31:827–832

Hsu PD, Lander ES, Zhang F (2014) Development and applications of CRISPR-Cas9 for genome engineering. Cell 157:1262–1278

Hu J, Lei Y, Wong WK, Liu S, Lee KC, He X, You W, Zhou R, Guo JT, Chen X, Peng X, Sun H, Huang H, Zhao H, Feng B (2014) Direct activation of human and mouse *Oct4* genes using engineered TALE and Cas9 transcription factors. Nucleic Acids Res 42:4375–4390

Hwang WY, Fu Y, Reyon D, Maeder ML, Tsai SQ, Sander JD, Peterson RT, Yeh JR, Joung JK (2013) Efficient genome editing in zebrafish using a CRISPR-Cas system. Nat Biotechnol 31:227–229

Ishino Y, Shinagawa H, Makino K, Amemura M, Nakata A (1987) Nucleotide sequence of the iap gene, responsible for alkaline phosphatase isozyme conversion in *Escherichia coli*, and identification of the gene product. J Bacteriol 169:5429–5433

Jansen R, Embden JD, Gaastra W, Schouls LM (2002) Identification of genes that are associated with DNA repeats in prokaryotes. Mol Microbiol 43:1565–1575

Jao LE, Wente SR, Chen W (2013) Efficient multiplex biallelic zebrafish genome editing using a CRISPR nuclease system. Proc Natl Acad Sci U S A 110:13904–13909

Jiang W, Bikard D, Cox D, Zhang F, Marraffini LA (2013) RNA-guided editing of bacterial genomes using CRISPR-Cas systems. Nat Biotechnol 31:233–239

Jinek M, Chylinski K, Fonfara I, Hauer M, Doudna JA, Charpentier E (2012) A programmable dual-RNA-guided DNA endonuclease in adaptive bacterial immunity. Science 337:816–821

Jinek M, East A, Cheng A, Lin S, Ma E, Doudna J (2013) RNA-programmed genome editing in human cells. Elife 2:e00471

Kearns NA, Genga RM, Enuameh MS, Garber M, Wolfe SA, Maehr R (2014) Cas9 effector-mediated regulation of transcription and differentiation in human pluripotent stem cells. Development 141:219–223

Kiani S, Beal J, Ebrahimkhani MR, Huh J, Hall RN, Xie Z, Li Y, Weiss R (2014) CRISPR transcriptional repression devices and layered circuits in mammalian cells. Nat Methods 11:723–726

Kim H, Kim JS (2014) A guide to genome engineering with programmable nucleases. Nat Rev Genet 15:321–334

Kim Y, Kweon J, Kim A, Chon JK, Yoo JY, Kim HJ, Kim S, Lee C, Jeong E, Chung E, Kim D, Lee MS, Go EM, Song HJ, Kim H, Cho N, Bang D, Kim S, Kim JS (2013) A library of TAL effector nucleases spanning the human genome. Nat Biotechnol 31:251–258

Kim JM, Kim D, Kim S, Kim JS (2014a) Genotyping with CRISPR-Cas-derived RNA-guided endonucleases. Nat Commun 5:3157

Kim S, Kim D, Cho SW, Kim J, Kim JS (2014b) Highly efficient RNA-guided genome editing in human cells via delivery of purified Cas9 ribonucleoproteins. Genome Res 24:1012–1019

Koike-Yusa H, Li Y, Tan EP, Velasco-Herrera Mdel C, Yusa K (2014) Genome-wide recessive genetic screening in mammalian cells with a lentiviral CRISPR-guide RNA library. Nat Biotechnol 32:267–273

Konermann S, Brigham MD, Trevino AE, Hsu PD, Heidenreich M, Cong L, Platt RJ, Scott DA, Church GM, Zhang F (2013) Optical control of mammalian endogenous transcription and epigenetic states. Nature 500:472–476

Kuscu C, Arslan S, Singh R, Thorpe J, Adli M (2014) Genome-wide analysis reveals characteristics of off-target sites bound by the Cas9 endonuclease. Nat Biotechnol. doi:10.1038/nbt.2916

Larson MH, Gilbert LA, Wang X, Lim WA, Weissman JS, Qi LS (2013) CRISPR interference (CRISPRi) for sequence-specific control of gene expression. Nat Protoc 8:2180–2196

Li JF, Norville JE, Aach J, McCormack M, Zhang D, Bush J, Church GM, Sheen J (2013a) Multiplex and homologous recombination-mediated genome editing in *Arabidopsis* and *Nicotiana benthamiana* using guide RNA and Cas9. Nat Biotechnol 31:688–691

Li W, Teng F, Li T, Zhou Q (2013b) Simultaneous generation and germline transmission of multiple gene mutations in rat using CRISPR-Cas systems. Nat Biotechnol 31:684–686

Li X, Burnight ER, Cooney AL, Malani N, Brady T, Sander JD, Staber J, Wheelan SJ, Joung JK, McCray PB Jr, Bushman FD, Sinn PL, Craig NL (2013c) *piggyBac* transposase tools for genome engineering. Proc Natl Acad Sci U S A 110:E2279–E2287

Lin Y, Cradick TJ, Brown MT, Deshmukh H, Ranjan P, Sarode N, Wile BM, Vertino PM, Stewart FJ, Bao G (2014) CRISPR/Cas9 systems have off-target activity with insertions or deletions between target DNA and guide RNA sequences. Nucleic Acids Res 42:7473–7485

Liu J, Gaj T, Patterson JT, Sirk SJ, Barbas CF 3rd (2014) Cell-penetrating peptide-mediated delivery of TALEN proteins via bioconjugation for genome engineering. PLoS One 9:e85755

Ma S, Chang J, Wang X, Liu Y, Zhang J, Lu W, Gao J, Shi R, Zhao P, Xia Q (2014) CRISPR/Cas9 mediated multiplex genome editing and heritable mutagenesis of *BmKu70* in *Bombyx mori*. Sci Rep 4:4489

Maeder ML, Linder SJ, Cascio VM, Fu Y, Ho QH, Joung JK (2013a) CRISPR RNA-guided activation of endogenous human genes. Nat Methods 10:977–979

Maeder ML, Angstman JF, Richardson ME, Linder SJ, Cascio VM, Tsai SQ, Ho QH, Sander JD, Reyon D, Bernstein BE, Costello JF, Wilkinson MF, Joung JK (2013b) Targeted DNA demethylation and activation of endogenous genes using programmable TALE-TET1 fusion proteins. Nat Biotechnol 31:1137–1142

Mali P, Yang L, Esvelt KM, Aach J, Guell M, DiCarlo JE, Norville JE, Church GM (2013a) RNA-guided human genome engineering via Cas9. Science 339:823–826

Mali P, Aach J, Stranges PB, Esvelt KM, Moosburner M, Kosuri S, Yang L, Church GM (2013b) CAS9 transcriptional activators for target specificity screening and paired nickases for cooperative genome engineering. Nat Biotechnol 31:833–838

Mali P, Esvelt KM, Church GM (2013c) Cas9 as a versatile tool for engineering biology. Nat Methods 10:957–963

Malina A, Mills JR, Cencic R, Yan Y, Fraser J, Schippers LM, Paquet M, Dostie J, Pelletier J (2013) Repurposing CRISPR/Cas9 for in situ functional assays. Genes Dev 27:2602–2614

Mashiko D, Fujihara Y, Satouh Y, Miyata H, Isotani A, Ikawa M (2013) Generation of mutant mice by pronuclear injection of circular plasmid expressing Cas9 and single guided RNA. Sci Rep 3:3355

Mashiko D, Young SA, Muto M, Kato H, Nozawa K, Ogawa M, Noda T, Kim YJ, Satouh Y, Fujihara Y, Ikawa M (2014) Feasibility for a large scale mouse mutagenesis by injecting CRISPR/Cas plasmid into zygotes. Dev Growth Differ 56:122–129

Matsunaga T, Yamashita JK (2014) Single-step generation of gene knockout-rescue system in pluripotent stem cells by promoter insertion with CRISPR/Cas9. Biochem Biophys Res Commun 444:158–163

Mendenhall EM, Williamson KE, Reyon D, Zou JY, Ram O, Joung JK, Bernstein BE (2013) Locus-specific editing of histone modifications at endogenous enhancers. Nat Biotechnol 31:1133–1136

Mercer AC, Gaj T, Fuller RP, Barbas CF 3rd (2012) Chimeric TALE recombinases with programmable DNA sequence specificity. Nucleic Acids Res 40:11163–11172

Mizuno S, Dinh TT, Kato K, Mizuno-Iijima S, Tanimoto Y, Daitoku Y, Hoshino Y, Ikawa M, Takahashi S, Sugiyama F, Yagami KI (2014) Simple generation of albino C57BL/6 J mice with G291T mutation in the tyrosinase gene by the CRISPR/Cas9 system. Mamm Genome 25:327–334

Mojica FJ, Díez-Villaseñor C, Soria E, Juez G (2000) Biological significance of a family of regularly spaced repeats in the genomes of archaea, bacteria and mitochondria. Mol Microbiol 36:244–246

Mojica FJ, Díez-Villaseñor C, García-Martínez J, Soria E (2005) Intervening sequences of regularly spaced prokaryotic repeats derive from foreign genetic elements. J Mol Evol 60: 174–182

Montague TG, Cruz JM, Gagnon JA, Church GM, Valen E (2014) CHOPCHOP: a CRISPR/Cas9 and TALEN web tool for genome editing. Nucleic Acids Res 42(Web Server issue):W401–407

Moore R, Chandrahas A, Bleris L (2014) Transcription activator-like effectors: a toolkit for synthetic biology. ACS Synth Biol. doi:10.1021/sb400137b

Nakagawa Y, Yamamoto T, Suzuki K, Araki K, Takeda N, Ohmuraya M, Sakuma T (2014) Screening methods to identify TALEN-mediated knockout mice. Exp Anim 63:79–84

Nakayama T, Fish MB, Fisher M, Oomen-Hajagos J, Thomsen GH, Grainger RM (2013) Simple and efficient CRISPR/Cas9-mediated targeted mutagenesis in *Xenopus tropicalis*. Genesis 51:835–843

Nekrasov V, Staskawicz B, Weigel D, Jones JD, Kamoun S (2013) Targeted mutagenesis in the model plant *Nicotiana benthamiana* using Cas9 RNA-guided endonuclease. Nat Biotechnol 31:691–693

Ni P, Zhang Q, Chen H, Chen L (2014) Inactivation of an integrated antibiotic resistance gene in mammalian cells to re-enable antibiotic selection. Biotechniques 56:198–201

Nishimasu H, Ran FA, Hsu PD, Konermann S, Shehata SI, Dohmae N, Ishitani R, Zhang F, Nureki O (2014) Crystal structure of Cas9 in complex with guide RNA and target DNA. Cell 156:935–949

Nissim L, Perli SD, Fridkin A, Perez-Pinera P, Lu TK (2014) Multiplexed and programmable regulation of gene networks with an integrated RNA and CRISPR/Cas toolkit in human cells. Mol Cell 54:698–710

Owens JB, Mauro D, Stoytchev I, Bhakta MS, Kim MS, Segal DJ, Moisyadi S (2013) Transcription activator like effector (TALE)-directed *piggyBac* transposition in human cells. Nucleic Acids Res 41:9197–9207

Park A, Won ST, Pentecost M, Bartkowski W, Lee B (2014) CRISPR/Cas9 allows efficient and complete knock-in of a destabilization domain-tagged essential protein in a human cell line, allowing rapid knockdown of protein function. PLoS One 9:e95101

Pattanayak V, Lin S, Guilinger JP, Ma E, Doudna JA, Liu DR (2013) High-throughput profiling of off-target DNA cleavage reveals RNA-programmed Cas9 nuclease specificity. Nat Biotechnol 31:839–843

Pennisi E (2013) The CRISPR craze. Science 341:833–836

Pourcel C, Salvignol G, Vergnaud G (2005) CRISPR elements in Yersinia pestis acquire new repeats by preferential uptake of bacteriophage DNA, and provide additional tools for evolutionary studies. Microbiology 151:653–663

Qi LS, Larson MH, Gilbert LA, Doudna JA, Weissman JS, Arkin AP, Lim WA (2013) Repurposing CRISPR as an RNA-guided platform for sequence-specific control of gene expression. Cell 152:1173–1183

Ramakrishna S, Kwaku Dad AB, Beloor J, Gopalappa R, Lee SK, Kim H (2014) Gene disruption by cell-penetrating peptide-mediated delivery of Cas9 protein and guide RNA. Genome Res 24:1020–1027

Ran FA, Hsu PD, Lin CY, Gootenberg JS, Konermann S, Trevino AE, Scott DA, Inoue A, Matoba S, Zhang Y, Zhang F (2013) Double nicking by RNA-guided CRISPR Cas9 for enhanced genome editing specificity. Cell 154:1380–1389

Reeks J, Naismith JH, White MF (2013) CRISPR interference: a structural perspective. Biochem J 453:155–166

Ronda C, Pedersen LE, Hansen HG, Kallehauge TB, Betenbaugh MJ, Nielsen AT, Kildegaard HF (2014) Accelerating genome editing in CHO cells using CRISPR Cas9 and CRISPy, a web-based target finding tool. Biotechnol Bioeng 111:1604–1616

Ru R, Yao Y, Yu S, Yin B, Xu W, Zhao S, Qin L, Chen X (2013) Targeted genome engineering in human induced pluripotent stem cells by penetrating TALENs. Cell Regeneration 2:5–12

Rusk N (2014) CRISPRs and epigenome editing. Nat Methods 11:28

Sakane Y, Sakuma T, Kashiwagi K, Kashiwagi A, Yamamoto T, Suzuki KT (2014) Targeted mutagenesis of multiple and paralogous genes in *Xenopus laevis* using two pairs of transcription activator-like effector nucleases. Dev Growth Differ 56:108–114

Sakuma T, Hosoi S, Woltjen K, Suzuki K, Kashiwagi K, Wada H, Ochiai H, Miyamoto T, Kawai N, Sasakura Y, Matsuura S, Okada Y, Kawahara A, Hayashi S, Yamamoto T (2013a) Efficient

TALEN construction and evaluation methods for human cell and animal applications. Genes Cells 18:315–326

Sakuma T, Ochiai H, Kaneko T, Mashimo T, Tokumasu D, Sakane Y, Suzuki K, Miyamoto T, Sakamoto N, Matsuura S, Yamamoto T (2013b) Repeating pattern of non-RVD variations in DNA-binding modules enhances TALEN activity. Sci Rep 3:3379

Sakuma T, Nishikawa A, Kume S, Chayama K, Yamamoto T (2014) Multiplex genome engineering in human cells using all-in-one CRISPR/Cas9 vector system. Sci Rep 4:5400

Sander JD, Joung JK (2014) CRISPR-Cas systems for editing, regulating and targeting genomes. Nat Biotechnol 32:347–355

Sander JD, Zaback P, Joung JK, Voytas DF, Dobbs D (2007) Zinc Finger Targeter (ZiFiT): an engineered zinc finger/target site design tool. Nucleic Acids Res 35(Web Server issue):W599–605

Sander JD, Maeder ML, Reyon D, Voytas DF, Joung JK, Dobbs D (2010) ZiFiT (Zinc Finger Targeter): an updated zinc finger engineering tool. Nucleic Acids Res 38(Web Server issue):W462–468

Shalem O, Sanjana NE, Hartenian E, Shi X, Scott DA, Mikkelsen TS, Heckl D, Ebert BL, Root DE, Doench JG, Zhang F (2014) Genome-scale CRISPR-Cas9 knockout screening in human cells. Science 343:84–87

Shan Q, Wang Y, Li J, Zhang Y, Chen K, Liang Z, Zhang K, Liu J, Xi JJ, Qiu JL, Gao C (2013) Targeted genome modification of crop plants using a CRISPR-Cas system. Nat Biotechnol 31:686–688

Shen B, Zhang W, Zhang J, Zhou J, Wang J, Chen L, Wang L, Hodgkins A, Iyer V, Huang X, Skarnes WC (2014) Efficient genome modification by CRISPR-Cas9 nickase with minimal off-target effects. Nat Methods 11:399–402

Smith C, Gore A, Yan W, Abalde-Atristain L, Li Z, He C, Wang Y, Brodsky RA, Zhang K, Cheng L, Ye Z (2014) Whole-genome sequencing analysis reveals high specificity of CRISPR/Cas9 and TALEN-based genome editing in human iPSCs. Cell Stem Cell 15:12–13

Sternberg SH, Redding S, Jinek M, Greene EC, Doudna JA (2014) DNA interrogation by the CRISPR RNA-guided endonuclease Cas9. Nature 507:62–67

Suzuki KT, Isoyama Y, Kashiwagi K, Sakuma T, Ochiai H, Sakamoto N, Furuno N, Kashiwagi A, Yamamoto T (2013) High efficiency TALENs enable F0 functional analysis by targeted gene disruption in *Xenopus laevis* embryos. Biol Open 2:448–452

Suzuki K, Yu C, Qu J, Li M, Yao X, Yuan T, Goebl A, Tang S, Ren R, Aizawa E, Zhang F, Xu X, Soligalla RD, Chen F, Kim J, Kim NY, Liao HK, Benner C, Esteban CR, Jin Y, Liu GH, Li Y, Izpisua Belmonte JC (2014) Targeted gene correction minimally impacts whole-genome mutational load in human-disease-specific induced pluripotent stem cell clones. Cell Stem Cell 15:31–36

Tsai SQ, Wyvekens N, Khayter C, Foden JA, Thapar V, Reyon D, Goodwin MJ, Aryee MJ, Joung JK (2014) Dimeric CRISPR RNA-guided FokI nucleases for highly specific genome editing. Nat Biotechnol 32:569–576

Veres A, Gosis BS, Ding Q, Collins R, Ragavendran A, Brand H, Erdin S, Talkowski ME, Musunuru K (2014) Low incidence of off-target mutations in individual CRISPR-Cas9 and TALEN targeted human stem cell clones detected by whole-genome sequencing. Cell Stem Cell 15:27–30

Walsh RM, Hochedlinger K (2013) A variant CRISPR-Cas9 system adds versatility to genome engineering. Proc Natl Acad Sci U S A 110:15514–15515

Wang H, Yang H, Shivalila CS, Dawlaty MM, Cheng AW, Zhang F, Jaenisch R (2013) One-step generation of mice carrying mutations in multiple genes by CRISPR/Cas-mediated genome engineering. Cell 153:910–918

Wang T, Wei JJ, Sabatini DM, Lander ES (2014) Genetic screens in human cells using the CRISPR-Cas9 system. Science 343:80–84

Westra ER, Buckling A, Fineran PC (2014) CRISPR-Cas systems: beyond adaptive immunity. Nat Rev Microbiol 12:317–326

Wiedenheft B, Sternberg SH, Doudna JA (2012) RNA-guided genetic silencing systems in bacteria and archaea. Nature 482:331–338

Wilkinson R, Wiedenheft B (2014) A CRISPR method for genome engineering. F1000Prime Rep 6:3

Wu X, Scott DA, Kriz AJ, Chiu AC, Hsu PD, Dadon DB, Cheng AW, Trevino AE, Konermann S, Chen S, Jaenisch R, Zhang F, Sharp PA (2014) Genome-wide binding of the CRISPR endonuclease Cas9 in mammalian cells. Nat Biotechnol. doi:10.1038/nbt.2889

Xiao A, Cheng Z, Kong L, Zhu Z, Lin S, Gao G, Zhang B (2014) CasOT: a genome-wide Cas9/gRNA off-target searching tool. Bioinformatics 30:1180–1182

Xie S, Shen B, Zhang C, Huang X, Zhang Y (2014) sgRNAcas9: A software package for designing CRISPR sgRNA and evaluating potential off-target cleavage sites. PLoS One 9:e100448

Yen ST, Zhang M, Deng JM, Usman SJ, Smith CN, Parker-Thornburg J, Swinton PG, Martin JF, Behringer RR (2014) Somatic mosaicism and allele complexity induced by CRISPR/Cas9 RNA injections in mouse zygotes. Dev Biol doi:. doi:10.1016/j.ydbio.2014.06.017

Yin H, Xue W, Chen S, Bogorad RL, Benedetti E, Grompe M, Koteliansky V, Sharp PA, Jacks T, Anderson DG (2014) Genome editing with Cas9 in adult mice corrects a disease mutation and phenotype. Nat Biotechnol 32:551–553

Yoshimi K, Kaneko T, Voigt B, Mashimo T (2014) Allele-specific genome editing and correction of disease-associated phenotypes in rats using the CRISPR-Cas platform. Nat Commun 5:4240

Yu C, Zhang Y, Yao S, Wei Y (2014) A PCR based protocol for detecting indel mutations induced by TALENs and CRISPR/Cas9 in zebrafish. PLoS One 9:e98282

Zhang Y, Ge X, Yang F, Zhang L, Zheng J, Tan X, Jin ZB, Qu J, Gu F (2014) Comparison of non-canonical PAMs for CRISPR/Cas9-mediated DNA cleavage in human cells. Sci Rep 4:5405

Zhao Y, Dai Z, Liang Y, Yin M, Ma K, He M, Ouyang H, Teng CB (2014) Sequence-specific inhibition of microRNA via CRISPR/CRISPRi system. Sci Rep 4:3943

Zhou Y, Zhu S, Cai C, Yuan P, Li C, Huang Y, Wei W (2014) High-throughput screening of a CRISPR/Cas9 library for functional genomics in human cells. Nature 509:487–491

Part II
Genome Editing in Cultured Cells and Various Organisms

Chapter 3
Editing Cultured Human Cells: From Cell Lines to iPS Cells

Hongmei Lisa Li and Akitsu Hotta

Abstract The human genome is complex, and the functions of many genes and specific genomic elements have not been fully explored. Cultured human cells, from conventional cell lines to iPS cells, are valuable tools that are used to study human genomics in the laboratory; however, the gene targeting efficiencies have generally been low in most of these human cells, making it difficult to assess the functions of these genes. The development of custom-engineered nucleases highlights the possibility of using different approaches to pursue functional genomic studies in cultured human cells. In this chapter, we review the chronological advances made in ZFNs, TALENs and CRISPR/Cas9 systems for gene disruption, locus-specific integration and gene correction. In addition, we summarize the emerging applications of tailor-made DNA binding modules for controlling the transcriptional state, modifying epigenetic marks and for live-cell imaging. The remaining challenges associated with increasing the target specificity, delivery and safety are also discussed.

Keywords CRISPR • ES (Embryonic Stem) cells • Gene correction • Gene therapy • Genome engineering • Human cells • iPS (induced pluripotent stem) cells • TALENs • ZFNs

3.1 Target cells

3.1.1 Human Cell Lines

Tissue culture of mammalian cells has been extensively used for biological studies, disease modeling and pharmacological testing. In general, primary cells may be isolated from a specific tissue directly, and these cells retain similar characteristics to the original cell type. However, primary cells have a limited capacity for proliferation, which eventually leads to cell senescence. On the other hand, cancer cells or transformed cells exhibit activation of oncogenes or genetic instability, which results in

H.L. Li • A. Hotta (✉)
Department of Reprogramming Science, Centre for iPS Cell Research and Application (CiRA), Kyoto University, Kyoto, Japan
e-mail: akitsu.hotta@cira.kyoto-u.ac.jp

T. Yamamoto (ed.), *Targeted Genome Editing Using Site-Specific Nucleases*, DOI 10.1007/978-4-431-55227-7_3

the acquisition of massive cell growth. Such transformed cells are often adaptable for in vitro cell culture, and once established as a cell line, they are able to grow indefinitely on tissue-culture dishes, which makes them as a valuable resource for biochemical and molecular studies. Since the first establishment of HeLa cells from cervical cancer in the early 1950s (Scherer et al. 1953), thousands of human cell lines derived from various tissues have been generated and are now available from public cell depositories, such as the ATCC (http://www.atcc.org).

3.1.2 Genome Editing in Cultured Mammalian Cells

Since the first success of homologous recombination in mammalian cell lines (Folger et al. 1982; Smithies et al. 1985), many attempts have been made using other mammalian cell lines. However, only a small subset of cell types, such as mouse ES cells (Thomas and Capecchi 1987) and chicken DT40 B lymphocytes (Buerstedde and Takeda 1991), showed relatively higher homologous recombination events compared with other cell lines.

3.1.3 ES/iPS Cells as an Attractive Target

ES cells are pluripotent cells that grow indefinitely while maintaining the ability to differentiate into all of the cell types that comprise the entire body. ES cells were first established from a mouse preimplantation embryo in 1981 (Evans and Kaufman 1981), and later from human blastocysts produced for IVF (in vitro fertilization) (Thomson et al. 1998). Remarkably, pluripotent stem cells exhibit the capacity for rapid proliferation while maintaining a normal karyotyping, although minor SNVs (single nucleotide variations) or CNVs (copy number variations) might occur during the culture.

To bypass the ethical problems involved in the use of human embryos, pluripotent stem cells can now be generated from somatic cells (i.e. skin fibroblasts or blood lymphocytes) by transient transduction of specific transcription factors, typically Oct3/4, Sox2, Klf4 and c-Myc (Takahashi and Yamanaka 2006; Takahashi et al. 2007).

The unique potential of mouse ES cells for contributing to a developing mouse embryo allows chimeric mice to be generated. If the ES cells are genetically modified, such as was first demonstrated following the integration of a retrovirus (Jaenisch 1977), the genetic modification also contributes to the chimeric mice and to germ cells at some frequency, which leads to the generation of genetically modified offspring. Such knockin or knockout mice are extremely useful to investigate the functions of a gene at an individual level, such as the presence of developmental defects and/or behavioral phenotypes (Capecchi 2001).

The approach of homologous targeting was applied to human ES cells. However, the results were less promising compared with those of mouse ES cell experiments, due to the fragile nature of human pluripotent stem cells when they were dissociated into single cells, and due to the relatively low frequency of transfection and homologous recombination. The first successful homologous recombination in human ES cells was reported in 2003 (Zwaka and Thomson 2003). Out of 350 drug-resistant clones analyzed, homologous recombination occurred in seven clones with attempted knockout of the HPRT locus. Another report identified that only two human ES cell clones out of 10^8 transfected cells had a disrupted HPRT gene induced by knocking-in a hygromycin resistance gene (Urbach et al. 2004).

The conventional knockin donor template normally uses a plasmid DNA with 1–20 kb homology arms, but BAC (bacterial artificial chromosome) DNA with longer homology arms (~100 kb) can be used as a donor template instead. In human ESCs, BAC-mediated homologous recombination has been reported to homozygously disrupt the ATM or p53 genes (Song et al. 2010), or knock-in GFP into the OSR1 gene (Mae et al. 2013). However, in general, transfected DNA predominantly integrates at random positions of chromosomes, and homologous recombination remains very rare.

3.2 The Development of Engineered Nucleases

Double strand breaks (DSBs) in genomic DNA are critical damage for all organisms, and must be repaired to maintain the integrity of the genomic information. From extensive DNA repair studies, it has been demonstrated that DSBs are repaired through homology-dependent recombination (HDR) or in a homology-independent manner, by so-called non-homologous end joining (NHEJ). To enhance the DNA repair pathways at desired sites, it is necessary to introduce a DSB at that site (Fig. 3.1).

3.2.1 Meganuclease (Homing Nuclease)

DSBs can be introduced at a defined site using a rare cutting homing endonuclease (also called a meganuclease, MN), I-Sce I, which recognizes about 18 bp DNA sequences (Rouet et al. 1994a, b). The site-specific induction of a DSB triggers the DNA repair pathways at the damaged site and enhances the homologous recombination (Choulika et al. 1995). Since I-SceI is derived from the mitochondrial DNA of yeast, there is no natural target site in the mouse or human genome that perfectly matches the target sequence of I-SceI (5′-TAGGGATAACAGGGTAAT-3′). Therefore, early studies had to integrate the target sequence into the host genome to test the meganuclease-mediated recombination within mammalian cells. Due to the limitations of the naturally existing meganucleases, custom engineering was required to target a desired locus.

To modify the target sequence of meganucleases, the DNA recognition domain has been engineered to alter their binding specificity (Epinat et al. 2003). Two meganucleases have been resolved to their protein structures by X-ray crystallography studies; one is I-CreI from chloroplast *Chlamydomonas reinhardtii* (PDB

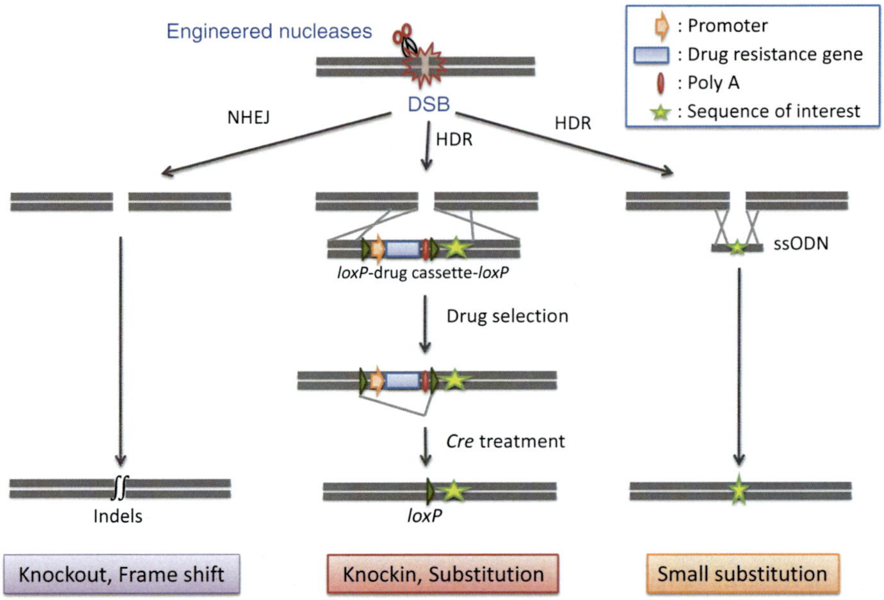

Fig. 3.1 Genome editing strategies used in cultured cells. Induction of DSB (double strand break) at the desired site by an engineered nuclease induces host DNA repair pathways, such as NHEJ (non-homologous end joining) or HDR (homology directed repair). Without donor template, NHEJ induces small indels, which can be used for gene knockout or frame-shifting. HDR is used to knockin or substitute the target sequence by using a donor template or ssODN (single stranded oligodeoxynucleotide)

code: 1G9Y (Chevalier et al. 2001)), and the other is I-DmoI from *Desulfurococcus mobilis* (PDB code: 1B24 (Silva et al. 1999)). Because the DNA-interacting amino acid residues in those meganucleases were known, randomized libraries of DNA binding residues were tested using a yeast screening system to identify the desired meganuclease to cleave at the desired target sequence (Chames et al. 2005; Grizot et al. 2010). With the engineered meganucleasess, endogenous RAG1 gene is targeted in 293H cells (Grizot et al. 2009), and Dystrophin gene is corrected in 293FT cells and human myoblasts (Chapdelaine et al. 2010). However, since the DNA binding domain of meganuclease is structurally associated with the DNA cleavage domain, engineering a meganuclease without compromising its overall protein stability was challenging.

3.2.2 ZFN (Zinc Finger Nuclease)

The zinc-finger DNA recognition domain is one of the most abundant DNA binding motifs in the mammalian genome, and has been manipulated to recognize a wide range of DNA sequences (Isalan et al. 1998; Dreier et al. 2001; Pabo et al. 2001).

The zinc-finger domain normally exists as an array of multiple domains, and each domain recognizes an approximately 3 bp DNA sequence. Extensive studies have revealed zinc-finger domains that can bind to all the possible 5′-GNN-3′ triplet DNA sequences (Segal et al. 1999; Dreier et al. 2000). With this knowledge, zinc-finger domains (typically three to four domains/a target sequence of nine or 12 bp) are conjugated with the nuclease domain from the FokI restriction enzyme, which specifically induces double-strand break only when it is dimerized (Smith et al. 2000; Bibikova et al. 2002).

The engineered chimeric proteins, named zinc-finger nucleases (ZFNs), offer much greater freedom for designing a target sequence compared with meganucleases. ZFNs have been applied to correct an X-linked severe combined immune deficiency (SCID) mutation in the IL2RG gene (Pro229) by homologous recombination in K562 myelogenous leukemia cells and human primary $CD4^+$ T cells (Urnov et al. 2005). In addition, ZFNs-mediated insertion of a hygromycin resistance donor template (knockin) has been tested in human iPS and ES cells to disrupt the endogenous PIG-A gene, which is required to retain glycosylphosphatidylinositol (GPI)-anchored proteins on the cell surface. Without ZFNs, the percentage of PIG-A-negative (knockout) cells was less than 0.2 % among hygromycin-resistant clones, which means that most of the hygromycin resistance cassettes were randomly integrated. However, when the ZFN is co-introduced with the donor template to target the PIG-A locus, more than 80 % of the PIG-A function was disrupted (Zou et al. 2009). In addition, the generation of knockin hES/iPS cells for targeting the OCT3/4, AAVS1 or PITX3 gene locus has been greatly facilitated by the use of zinc-finger nucleases (Hockemeyer et al. 2009). The tables below provide examples of ZFN-mediated genome editing technologies applied in human cell lines (Table 3.1), pluripotent stem cells (Table 3.2) and other cell types including primary cells (Table 3.3).

Table 3.1 Examples of ZFN-mediated genome editing in cultured human cell lines

Target cells	Gene editing	Target gene	References
K562	Knockin	*IL2RG*	Urnov et al. (2005)
K562, CEM, U937, Jurkat	Knockin	*IL2RG, CCR5*	Lombardo et al. (2007)
293T, K562	Disruption, Knockin	*VEGFA, HOXB13, CFTR, IL2Rg*	Maeder et al. (2008)
293T	Disruption	*CCR5*	Kim et al. (2009)
293T, K562, Hep3B, U2OS	Disruption, Knockin	*AAVS1*	DeKelver et al. (2010)
293T	Disruption	*DMD*	Rousseau et al. (2011)
293T, K562	Knockin	*IL2RG, CCR5*	Lombardo et al. (2011)
K562, Hep3B	Knockin	*F9*	Li et al. (2011)
Jurkat	Disruption	*TCR*	Provasi et al. (2012)
K562	Knockin	*AAVS1*	Chang and Bouhassira (2012)
HT1080	Knockin	*DYRK1A*	Jiang et al. (2013a)

Table 3.2 Examples of ZFN-mediated genome editing in human pluripotent stem cells

Target cells	Gene editing	Target gene	References
ES cells, iPS cells	Knockin	*AAVS1, OCT3/4, PITX3*	Hockemeyer et al. (2009)
ES cells, iPS cells	Knockin	*PIG-A*	Zou et al. (2009)
ES cells	Knockin	*AAVS1*	DeKelver et al. (2010)
ES cells, iPS cells	Knockin	*AAVS1, OCT3/4, PITX3*	Hockemeyer et al. (2011)
ES cells, iPS cells	Disruption	*CCR5*	Yao et al. (2011)
ES cells	Knockin	*CCR5*	Lombardo et al. 2007)
ES cells, iPS cells	Knockin	*α-Synuclein*	Soldner et al. (2011)
iPS cells	Knockin	*AAVS1*	Zou et al. (2011b)
iPS cells	Knockin	*β-Globin*	Zou et al. (2011a)
iPS cells	Knockin	*α1-Antitrypsin*	Yusa et al. (2011)
iPS cells	Knockin	*β-Globin*	Sebastiano et al. (2011)
iPS cells	Knockin	*AAVS1*	Chang and Bouhassira (2012)
iPS cells	Knockin	*DYRK1A* on chr 21	Jiang et al. (2013a)

Table 3.3 Examples of ZFN-mediated genome editing in other human cells

Target cells	Gene editing	Target gene	References
Primary $CD4^+$ T cells	Disruption, Knockin	*IL2Rg*	Urnov et al. (2005)
Primary $CD4^+$ T cells	Disruption	*CCR5*	Perez et al. (2008)
$CD34^+$ hematopoietic stem/progenitor cells	Disruption	*CCR5*	Holt et al. (2010)
Myoblasts	Disruption	*DMD*	Rousseau et al. (2011)
Lymphoblastoid cells, $CD34^+$ progenitor cells	Knockin	*IL2Rg, CCR5, AAVS1*	Lombardo et al. (2007)
Primary T lymphocytes	Disruption	*TCR*	Provasi et al. (2012)
Primary T lymphocytes	Disruption	*TCR*	Torikai et al. (2012)

Hence, ZFNs represent a powerful tool to modify genomic sequences in cultured human cells; however, their bias for a higher GC-content of target sequence (Kim et al. 2009), the unavailability of some triplet sequences (Segal et al. 1999; Bae et al. 2003), cytotoxicity owing to off-target effects (Cornu et al. 2008), context-dependent activity and patent restrictions have compromised the use of ZFNs by most scientists.

3.2.3 TALEN

A novel DNA binding domain was discovered from a plant pathogen, *Xanthomonas*, which is known as transcription activator-like effector (TAL effector or TALE). The DNA binding domain of TALEs consists of 10 to 30 tandem repeats of RVD (repeat variable diresidue) domains. Each RVD domain recognizes a single base-pair of DNA based on two amino acid residues, such as NI for A, HD for C, NG for T and NN for G/A (Boch et al. 2009; Moscou and Bogdanove 2009). Such a straightforward DNA decipher code simplifies the assembly of customized nucleases (Christian et al. 2010; Hockemeyer et al. 2011). By joining the custom-engineered TALE DNA binding domain with a FokI nuclease domain, a more flexible design of the target sequence becomes feasible. With this TALE nuclease (TALEN) system, the human endogenous NTF3 and CCR5 genes were edited in the human K562 cell line (Miller et al. 2011) and other cell lines (Table 3.4). The effectiveness of TALENs has also been demonstrated in human ES/iPS cells (Hockemeyer et al. 2011; Ding et al. 2013a) (Table 3.5) and other cell types (Table 3.6).

Table 3.4 Examples of TALEN-mediated genome editing in cultured human cell lines

Target cells	Gene editing	Target gene	References
293T	Disruption, Knockin	*CCR5, IL2RG*	Mussolino et al. (2011)
293T, K562	Disruption, Knockin	*NTF3, CCR5*	Miller et al. (2011)
293T	Disruption	*HPRT1*	Cermak et al. (2011)
U2OS	Disruption	*AXIN2, BRCA1, CDC73, CHD7, CYLD, ERCC2, JAK2, MYCN, NCOR2, NBN, XPC*	Reyon et al. (2012)

Table 3.5 Examples of TALEN-mediated genome editing in human pluripotent stem cells

Target cells	Gene editing	Target gene	References
ES cells, iPS cells	Knockin	*AAVS1, OCT3/4, PITX3*	Hockemeyer et al. (2011)
ES cells, iPS cells	Disruption	*AKT2, ANGPTL3, APOB, ATGL, C6OR106, CELSR2, GLUT4, LINC00116, PLIN1, SORT1, TRIB1, CIITA, CFTR, NLRC5*	Ding et al. (2013a)
iPS cells	Disruption	*HPRT*	Sakuma et al. (2013)
iPS cells	Knockin	*α1-Antitrypsin*	Choi et al. (2013)
iPS cells	Knockin	*COL7A1*	Osborn et al. (2013)
iPS cells	Knockin	*β-Globin*	Ma et al. (2014)

Table 3.6 Examples of TALEN-mediated genome editing in other human cells

Target cells	Gene editing	Target gene	References
Immortalized myoblasts	Disruption	*DMD*	Ousterout et al. (2013)
Fusion of dermal fibroblasts and osteosarcoma cell line	Disruption, Knockin	*Mitochondrial DNA, MT-ND6*	Bacman et al. (2013)
Primary fibroblasts	Knockin	*COL7AI*	Osborn et al. (2013)
Primary T cells	Disruption	*TCR*	Boissel et al. (2013)

3.2.4 CRISPR/Cas9

In bacteria and archaea, an RNA-mediated adaptive defense mechanism, known as the clustered regularly interspaced short palindromic repeats (CRISPR) system, exists to fight off plasmid or bacteriophage invasions. The Type II CRISPR system is unique in that the CRISPR-associated 9 (Cas9) protein induces DSB in bacteria when associated with CRISPR RNA (crRNA) and trans-activating crRNA (tracrRNA). The system is simplified by conjugation of crRNA and tracrRNA into a single guide RNA (sgRNA) (Jinek et al. 2012). Remarkably, the RNA component of the CRISPR system determines the target sequence based on the Watson-Crick base pair rules. Therefore, the design and construction of a target-specific sgRNA is versatile and straightforward. To adapt the prokaryotic system to work in eukaryotic cells, several groups have attached a NLS (nuclear localization signal) to the Cas9 protein, and have demonstrated the efficient induction of DSBs in HEK293T cells (Jinek et al. 2013), mouse neuro2A (N2A) cells (Cong et al. 2013), K562 cells and human ES/iPS cells (Mali et al. 2013b; Ding et al. 2013b). Here, we summarize the emerging applications of the CRISPR/Cas9 system in human cell lines (Table 3.7), pluripotent stem cells (Table .3.8) and other primary cells (Table 3.9).

Table 3.7 Examples of CRISPR/Cas9-mediated genome editing in cultured human cell lines

Target cells	Gene editing	Target gene	References
293T, K562	Disruption, Knockin	*AAVS1, DNMT3A/3B*	Mali et al. (2013b)
293T	Disruption, Knockin	*EMX1, PVALB*	Cong et al. (2013)
293T	Disruption	*CATL1*	Jinek et al. (2013)
293T	Disruption	*β-Globin, CCR5*	Cradick et al. (2013)
K562	Disruption	*CCR5, C4BPB*	Cho et al. (2013)
K562	Disruption	*CCR5*	Cradick et al. (2013)
293T, HeLa, Jurkat	Disruption	*HIV-1 LTR*	Ebina et al. (2013)
293T	Disruption	*EMX1, PVALB*	Hsu et al. (2013)
293T	Disruption	*β-Globin, CCR5*	Lin et al. (2014)
293T, K562, HeLa, TCT116, A549	Disruption	*C4BPB, HLA-B, CTNNB1, KRAS, PIK3CA, IDH1*	Kim et al. (2014)
293T	Disruption, Knockin (*1)	*EMX1, VEGFA, DYRK1A*	Ran et al. (2013)
293T, HeLa	Disruption (*1)	*RAG1, VEGFA, EMX1*	Shen et al. (2014)
U2OS	Disruption (*2)	*VEGFA, EMX1, CLTA*	Fu et al. (2014)
293 T, U2OS	Disruption (*3)	*APC, BRCA1, EMX1, FANCF, GLI1, MLH1, RARA, SS18, VEGFA*	Tsai et al. (2014)
293 T	Disruption	*HPRT1, ATM, APC, CDH1, AXIN2, CFTR*	Sakuma et al. (2014)

*1: mediated by double nicking approach, which uses two adjacent sgRNAs and Cas9 nickase (D10A mutant)

*2: mediated by short truncated sgRNA (tru-RGNs)

*3: mediated by catalytically inactive Cas9 (dCas9) fused with FokI nuclease domain

Table 3.8 Examples of CRISPR/Cas9-mediated genome editing in human pluripotent stem cells

Target cells	Gene editing	Target gene	References
iPS cells	Knockin	*AAVS1*	Mali et al. (2013b)
ES cells, iPS cells	Disruption, Knockin	*AKT2, CELSR2, GLUT4, LDLR, LINC00116, SORT1,*	Ding et al. (2013b)
ES cells	Knockin (*1)	*EMX1*	Ran et al. (2013)
ES cells, iPS cells	Disruption, Knockin	*NGN3, GATA4, GATA6, TET1, TET2, TET3, APOE*	Gonzalez et al. (2014)
iPS cells	Disruption	*CCR5*	Ye et al. (2014)

*1: mediated by double nicking approach, which uses two adjacent sgRNAs and Cas9 nickase (D10A mutant)

Table 3.9 Examples of CRISPR/Cas9-mediated genome editing in primary human cells

Target cells	Gene editing	Target gene	References
$CD4^+$ T cells	Disruption	*CCR5*	Ebina et al. (2013)
Adult intestinal stem cells, primary cultured small intestinal and large intestinal stem cells	Disruption, Knockin	*APC, CFTR*	Schwank et al. (2013)

3.3 Delivery of Engineered Nucleases

To achieve successful gene targeting in the desired cell type, it is critical to establish a method to introduce the nuclease component into the target cells. Nucleases can be delivered as DNA, mRNA or protein, but DNA transfection is the most common choice for basic research purposes. In general, there are three categories of DNA transfection methods used to deliver a nuclease-coding nucleic element into mammalian cells: biochemical methods, physical methods and viral methods.

3.3.1 Chemical Transfection

Biochemical methods use chemical reagents to deliver DNA into cells. Several lipofection reagents are now available from many manufactures, such as Lipofectamine 2000 from Life Technologies or FuGENE HD from Roche (now available from Promega). These lipofection reagents are very efficient and highly scalable for many transfection conditions, making them a first choice for testing and screening several nucleases. Ideally, it is better to compare the efficiencies of different transfection reagents with the target cells in advance with a transfection control vector.

3.3.2 Physical Transfection

Physical methods, such as electroporation or nucleofection, are used to transfect hard-to-transfect cells, such as primary cells or stem cells. The original electroporation method was developed in the early 1980s (Neumann et al. 1982), and used electrical pulses to create transient pores in the cell membrane which allowed foreign DNA to be introduced. Electroporation has been widely applied from prokaryotic to eukaryotic cells, but the electrical parameters, such as the voltage and/or pulse width, must be pre-determined in order for efficient transduction to be achieved. To increase the efficiency of introducing DNA into the nucleus, a nucleofection method was developed by Amaxa in 1998 using a combination of high-voltage pulses and a high buffer/ionic strength solution [US patent US 7,332,332 B2]. The neon transfection system (initially called a Microporator) was developed by Digital Bio in 2006, and uses a pipette-type electrode instead of a standard electroporation cuvette (Kim et al. 2008). A Japanese biotech company, Nepagene, developed a novel electroporator called NEPA21, which makes use of a combination of two types of electrical pluses: a poring pulse and a transfer pulse. The first short pulses with high voltage make pores in the cell membrane, while the second long pulses with low voltage introduce DNA inside the cells.

3.3.3 Viral Transfection

Viral vectors are very efficient tools to introduce foreign genes into mammalian cells, as they use the virus' natural entry machinery. For example, an adeno-associated virus (AAV) vector was used to deliver ZFNs into mouse cells (Li et al. 2011), adenoviral vectors were used to introduce TALENs (Holkers et al. 2013) or CRISPR/Cas9 (Maggio et al. 2014) and baculovirus was used to deliver ZFNs (Phang et al. 2013; Tay et al. 2013) or TALENs (Zhu et al. 2013). An integrase-deficient lentivirus (IDLV) vector was used to deliver ZFNs and donor template DNA (Lombardo et al. 2007), but the TALEN is difficult to deliver owing to the repeated sequence of its DNA binding domain (Holkers et al. 2013). Lentiviral vectors have been used to deliver Cas9 and sgRNAs (Shalem et al. 2014; Wang et al. 2014; Koike-Yusa et al. 2014).

3.3.4 Bacterial Transfection

Some pathogenic bacteria, such as *Pseudomonas aeruginosa,* have a needle-like appendage to inject bacterial proteins into host cells. For bacteria-mediated transfection, the targeting TALEN protein is expressed in *Pseudomonas aeruginosa*, then it is infected into HeLa cells by *Pseudomonas aeruginosa*. As a result, the TALEN proteins are directly injected into HeLa cells via the type III secretion system (T3SS) (Jia et al. 2014).

3.3.5 *Protein Transfection*

The successful delivery of a purified Cas9 protein produced by *E. coli* was also demonstrated. The purified Cas9 protein from *E. coli* was chemically conjugated with a cell-penetrating peptide (4-maleimidobutyryl- GGGRRRRRRRRRLLLL). By simply adding the conjugated Cas9 protein and sgRNA into culture media (Ramakrishna et al. 2014), or electroporation by an Amaxa Nucleofecter (Kim et al. 2014), the endogenous CCR5 gene was successfully disrupted in several cell types, such as HEK293T, HeLa, NCCIT human embryonic carcinoma cells, H9 hESCs and dermal fibroblasts.

3.4 Genomic Manipulation Methods and Their Applications

3.4.1 *Gene Knockout*

The human genome project has identified nearly 18,000 protein coding genes in the human genome. However, the functions of many genes are still unclear. Mutating a gene to destroy its function, so called gene knockout or loss-of-function, is an essential technique used to study the function of a gene. Knockout methods have been extensively applied to ES cells to generate mouse models of disease or to investigate the function of a gene during mouse development. The engineered nuclease technologies now open the door for gene knockout studies that are more accessible for various types of human cells, as described in this chapter, but also for other non-model organisms, as described in other chapters.

3.4.2 *Genome-Wide Loss-of-Function Libraries*

Combinatorial synthesis of TALENs or sgRNA allows genome wide libraries to be constructed that can target a large panel of mammalian genes. Below, we summarize the genome-wide loss-of-function libraries developed based on the TALEN (Table 3.10) or CRISPR (Table 3.11) systems. Such libraries will aid in the study of gene functions via functional genomics in human cells.

Table 3.10 The TALEN-based genome-wide libraries

Library size	Target genes	Tested cells	Functional assay	References
540	274 miRNAs	SK-BR-3 human breast cancer cells	Knockout miRNAs / cell proliferation, changes in the global mRNA levels	Kim et al. (2013b)
169,362	18,740 genes	HEK293T cells	Disruption of NF-κB signaling pathways	Kim et al. (2013a)

Table 3.11 CRISPR/Cas9-based genome-wide libraries

Library size	Target genes	Tested cells	Functional assay	References
64,751	18,080 genes	A375 cells / hESCs HUES2	Resistance to vemurafenib	Shalem et al. (2014)
73,151	7,114 genes	KBM7 cells / HL-60 cells	Resistance to 6-thioguanine	Wang et al. (2014)
87,897	19,150 genes	Mouse JM8 ESCs	Resistance to either *Clostridium septicum* alpha-toxin or 6-thioguanine	Koike-Yusa et al. (2014)
869	291 genes	HeLa cells	Resistance to diphtheria-toxin or chimeric anthrax toxin	Zhou et al. (2014)

3.4.3 Reporter Cell Lines

Each gene has an expression pattern that depends on the cell types or developmental stage. Visualization of the spatiotemporal regulation of a gene is useful to track cells, to optimize the treatment conditions or to enrich the target cells of interest, such as emerging iPS cell colonies (Hotta et al. 2009). To achieve these goals, a reporter gene, such as GFP, can be inserted at the translational start site or immediately after the gene (before the poly A signal) connected by an IRES (internal ribosome entry site) or a 2A self-cleavage peptide. This way, the reporter gene will be expressed, reflecting the endogenous expression pattern of the targeted gene. In mouse experiments, lineage tracing or isolation of particular cell types is feasible by reporter knockin techniques. In human pluripotent stem cells, such reporter knockin cell lines are useful to monitor the state of pluripotency or differentiation. For example, pluripotent stem cell-specific genes, OCT3/4 (Zwaka and Thomson 2003) and REX1 (Bhatia et al. 2013), have been targeted for GFP knockin to monitor the pluripotent state of living human ES cells.

The differentiation of human pluripotent stem cells towards a particular cell lineage of interest is a critical step in stem cell field; however, it requires extensive and laborious optimization of the differentiation conditions. To aid in the optimization process, several lineage-specific genes have been targeted, such as MIXL1, to mark primitive streak-like cells (Davis et al. 2008), OLIG2 for neuroglial cells, FEZF2 (FEZL) for corticospinal neurons (Ruby and Zheng 2009) or OSR1 for the intermediate mesoderm (Mae et al. 2013).

3.4.4 Knockin Transgenic Studies

Transgenes can be inserted into a targeted locus to achieve consistent expression under a defined copy number of the transgene. For this purpose, the Rosa26 locus or AAVS1 (also known as the PPP1R12C gene) locus have been widely used due to their "open" chromatin nature in essentially all cell types (Lombardo et al. 2011). For example, the expression of GFP from the Rosa26 or AAVS1 locus under the control of a strong ubiquitous promoter allowed cell tracking experiments (Wang et al. 2012) . When a doxycycline-inducible expression cassette was inserted into the AAVS1 intron 1 locus by ZFNs, such hESCs expressed EGFP dependent on the doxycycline concentration (Hockemeyer et al. 2009). The expression of GFP from the AAVS1 locus can be tightly regulated not only in undifferentiated hiPSCs, but also in differentiated cells in vitro and in vivo (Qian et al. 2014). The AAVS1 locus has also been targeted in other cell types, such as epithelial stem cells from the epidermis (Coluccio et al. 2013), mesenchymal stem cells (MSCs) (Li et al. 2013) and hematopoietic stem cells from cord blood (Genovese et al. 2014).

3.4.5 Knockin Gene Therapy

In addition to a reporter construct, a therapeutic gene for gene therapy can be inserted into the AAVS1 locus, as AAVS1 is considered to be one of the "safe-harbors" that can insert a foreign gene without disruption of the normal cell homeostasis. There are several reports of studies for this purpose, such as insertion of the α-globin gene for α-thalassemia (Chang and Bouhassira 2012), the FANCA gene for Fanconi anemia (Río et al. 2014) and RPS19 (Ribosomal protein S19) for Damond Blackfan anemia patients (Garçon et al. 2013).

The knockin approach was also used to target the transcriptional start site of coagulation factor 9 (F9) to treat hemophilia B model mice. By introducing the F9 cDNA sequence into the transcriptional start site, the cDNA could be expressed from its endogenous promoter. In addition, this approach can be applied for many patients, regardless of the mutation type within F9 gene (Li et al. 2011).

In females, one of the X chromosomes is inactivated to maintain the dosage of X-linked gene transcripts. A long non-coding RNA, Xist, on X chromosomes plays a central role to inactivate the second X chromosome. The insertion of Xist into one of the trisomy 21 chromosomes (at DYRK1A locus) by ZFNs has been investigated to epigenetically silence one copy of chromosome 21, which could potentially be used to treat Down syndrome (Jiang et al. 2013b).

3.4.6 Chromosomal Deletion or Inversion

Treatment with single nucleases normally results in small deletions, ranging from a few base pairs to 100 base pairs. By using two pairs of ZFNs or TALENs, deletion of an up to 230 kbp region, 15 kbp duplication or inversion of up to 140 kbp was possible in HEL293T cells (Lee et al. 2012). Such inversion of a large genomic region was demonstrated in the blood coagulation Factor VIII (F8) gene to model a mutation associated with hemophilia A in human iPS cells (Park et al. 2014).

3.5 Other Applications of Custom DNA Binding Modules

Apart from genome engineering, custom DNA binding modules have been utilized for various applications in cell biological studies, as shown in Fig. 3.2.

3.5.1 Controlling the Transcriptional Status

The zinc-finger domain and TAL effector domain are naturally conjugated with transcriptional activator or suppressor domains. By modulating their targeting specificity, one can generate an "artificial transcription factor" to target a gene of interest to modulate its transcription (Blancafort et al. 2004). For instance, the activation domain of the VP16 protein from herpes simplex virus is able to recruit TBP (TATA-binding protein) and other transcriptional initiation complex components to drive transcription when fused to a DNA binding domain (Sadowski et al. 1988).

In this context, TALE was conjugated with a VP16 domain to activate the endogenous human NTF3 gene in HEK293 cells (Miller et al. 2011). To enhance the activation efficacy, the TALE domain was also fused with a strong NF-κB p64

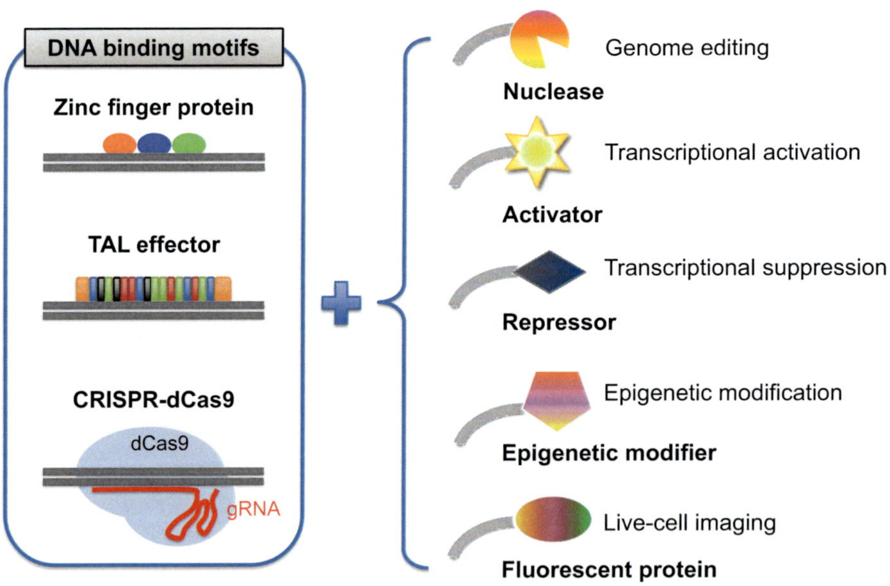

Fig. 3.2 Modular assembly of functional domains for various applications DNA binding motifs, such as zinc finger domain, TAL effector, or catalytically inactive Cas9 (dCas9)-sgRNA, can be used to recruit a functional domain of interest at the desired site(s). Apart from the Fok I nuclease domain, several transcriptional activator or suppressor domains have been conjugated to modulate transcription of a target gene

transcriptional activation domain, or VP64, which is composed of four tandem repeats of the VP16 core activation domain. The TALE-VP64 fusion protein was introduced into HEK293 cells and BJ human fibroblasts to increase the level of expression of VEGFA and the miR-302/367 cluster, respectively (Maeder et al. 2013b).

The DNA binding domain of TALE has been replaced with sgRNA and catalytically dead dCas9 protein (normally D10A and H840A mutations). The C-terminus of dCas9 was conjugated with VP64 to activate a GFP reporter target, or KRAB suppressor domain, to suppress the GFP reporter or endogenous CD71 and CXCR4 genes in HeLa cells (Gilbert et al. 2013). The activation of endogenous human genes by dCas9-VP64 fusion proteins has also been investigated in HEK293 cells with various target genes, such as the ASCL1, NANOG, HBG1, HBG2 and MYOD1 genes (Perez-Pinera et al. 2013), as well as the VEGFA and NTF3 genes (Maeder et al. 2013a) or IL1RN and SOX2 genes (Cheng et al. 2013). The dCas9-VP64 fusion protein can be used to monitor the binding specificity of sgRNA and Cas9. A target sequence with various mutations can be targeted with dCas9-VP64 to transcribe a 24-bp barcode and tdTomato reporter, and an acceptable number of mutations can indicate the target specificity of the sgRNA (Mali et al. 2013a). The VP64 activation domain can be bound to sgRNA via the MS2 protein and its binding RNA sequence on the MS2 stem loop (Mali et al. 2013a).

The DNA binding activity of gRNA-Cas9 itself can be used to block transcriptional initiation or elongation by targeting the transcriptional start site in *E. coli*, by the so-called CRISPRi (Qi et al. 2013), but the suppression was modest in HEK293 cells compared with dCas9-KRAB constructs.

3.5.2 Modifying Epigenetic Marks

As an analog approach to fusing a transcriptional activator/suppressor domain to recruit an epigenetic modifier to the target site to alter the transcriptional status, fusion of an epigenetic modifier directly to the DNA binding domain has been investigated to change the epigenetic status, such as the DNA methylation or histone modification (Voigt and Reinberg 2013; Rusk 2014).

For example, conjugation of the catalytic domain of TET1 hydroxymethyltransferase with the TALE DNA binding domain resulted in a local reduction of the CpG methylation level. When intron 2 of the KLF4 gene was targeted in K562 cells, a 21–30 % reduction in the CpG methylation level was observed 10 and 16 bp away from the TALE binding site. The promoter regions of the RHOXF2 and β-globin (HBB) genes were targeted, and mRNA activation and local demethylation were observed (Maeder et al. 2013a). Conjugation of the TALE domain with LSD1 histone demethylase, which catalyzed the removal of H3K4 methylation, was tested to target an enhancer region of the SCL (stem cell leukemia, TAL1 gene) locus. The introduction of the TALE-LSD1 fusion protein into K562 cells resulted in a reduction of H3K4me2 near the TALE binding site, and induced the downregulation of the proximal genes (Mendenhall et al. 2013).

3.5.3 *Optical Control of Transcription*

As a sophisticated example of artificially controlling transcription, the TALE domain has been fused with the light-sensitive cryptochrome 2 (CRY2) protein from *Arabidopsis thaliana*, in the LITEs (Light-inducible transcriptional effectors) technique. When CRY2 is illuminated with blue light, it changes its conformation and recruits its interacting partner, CIB1. By conjugating CIB1 with the VP64 activation domain, the target genes with TALE binding sites are successfully upregulated only when 466 nm blue light is applied. These included the Neurog2 gene in primary mouse neurons, and the Grm2 gene in living mouse brain. In addition, conjugation of histone effector domains with the CRY2-CIB1 system was demonstrated to alter the mRNA transcriptional status of a target gene (Grm2) in primary neurons, and such histone modifiers include histone methyltransferases (KYP, SET8, NUE, and PHF19) and histone deacetylases (Sin3a, Sirt3, NcoR, HDAC8, RPD3 and Sir2a) (Konermann et al. 2013).

3.5.4 *Visualization for Live-Cell Imaging*

The flexible design of the DNA binding domains allows for anchoring a fluorescent protein into the desired location of the genome in living cells. For example, the TALE DNA binding domain was fused with the mClover fluorescent protein to target major satellite sequences (pericentromeric region) in mouse ES cells (Miyanari et al. 2013). Similarly, a TALE domain was conjugated with Venus or mCherry to label telomeric or centromeric repeats in human osteosarcoma U2OS cells (Ma et al. 2013). Catalytically inactive Cas9 (dCas9) was also fused with GFP to target the telomere region by a telomere-sequence specific sgRNA in HEK293, UMUC3 and HeLa cells (Chen et al. 2013).

At present, only repetitive sequence regions (telomere or centromere) can be targeted to obtain sufficient fluorescent signals, as there is no way to discriminate DNA binding fluorescent protein from free-floating fluorescent protein, which results in a high background. To visualize a non-repetitive single-copy gene as a spot, it is reported that at least 26 probes (26 sgRNAs and dCas9-GFP) need to be targeted within a 5 kb window (Chen et al. 2013). One might be able to visualize with higher resolution with better signal-to-noise ratio by using a combination of highly sensitive single-molecule imaging techniques and fluorescence resonance energy transfer (FRET) technology to detect the properly binding fluorescence probes at the desired sites with proper orientation.

3.5.5 *Pull-Down of a Certain DNA Region by enChIP*

Chromatin immunoprecipitation (ChIP) experiments use a DNA binding protein to pull-down its target DNA sequences. By using a custom DNA binding protein, such as TALE or dCas9-sgRNA, the desired sequence region can be isolated, via the "engineered DNA-binding molecule-mediated chromatin immunoprecipitation (enChIP)" technique. A TALE domain with a NLS and 3xFLAG tag was used to target the telomeric region in mouse hematopoietic cells Ba/F3, and pull-down by an anti-FLAG antibody identified telomere DNA by Southern blotting, telomere binding proteins by a mass spectrometry analysis and telomere-associated RNA by RT-PCR (Fujita et al. 2013). A similar approach was possible using dCas9 fused with a FLAG tag (Fujita and Fujii 2013).

3.6 Future Perspectives for Gene Therapy

In the field of gene therapy research, it has long been a desire to correct genetic mutations to cure genetic diseases. The development of more efficient genome editing tools has opened the possibility of precise gene correction in such patients (Lisa Li et al. 2014).

One unique gene therapy approach employing an engineered nuclease was developed to treat HIV by disruption of the CCR5 gene. CCR5 is used as a co-receptor for HIV to infect CD4+ T cells. Therefore, disruption of CCR5 in T cells or progenitor cells would lead to resistance against HIV (Perez et al. 2008; Holt et al. 2010; Maier et al. 2013). Phase I clinical trials were performed to treat $CD4^+$ T cells from 12 HIV patients by adenoviral delivery of ZFNs targeting the CCR5 gene, and the procedures appeared to be safe and tolerated during the evaluation period (36 weeks) (Tebas et al. 2014). Encouragingly, the HIV genomic RNA level became undetectable in one of the treated patients, but the patient later appeared to carry a natural deletion of the CCR5 gene (delta 32) in one allele. This emphasizes the importance of biallelic disruption of the CCR5 gene to achieve a successful treatment. Further studies will reveal the feasibility and effectiveness of this gene knockout strategy for the treatment of HIV/AIDS.

Conclusions

The rapid development of genome engineering technologies, such as the ZFN, TALEN and CRISPR/Cas9 systems, have broadened our abilities to modify specific locations of genomic sequences in human cells. The applications of such genome engineering techniques are broad, and range from gene function studies to disease modeling to the generation of reporter lines to gene correction (gene therapy). Conjugation of transcriptional regulators, epigenetic modifiers or fluorescent proteins will improve the ability to explore gene regulation, the regulation of the local epigenetic status and to perform live-cell imaging of certain genomic regions. Challenges still remain, particularly with regard to increasing the binding specificity and affinity, and to expand the flexibility of the target design. The temporal and spatial regulation of the DNA binding kinetics is another important area to be explored. These novel tools have the potential to alter experimental biology using cells, including human cells, and should expand our understanding of biology and disease.

References

Bacman SR, Williams SL, Pinto M, Peralta S, Moraes CT (2013) Specific elimination of mutant mitochondrial genomes in patient-derived cells by mitoTALENs. Nat Med 19:1111–1113

Bae KH, Kwon YD, Shin HC, Hwang MS, Ryu EH, Park KS, Yang HY, Lee DK, Lee Y, Park J, Kwon HS, Kim HW, Yeh BI, Lee HW, Sohn SH, Yoon J, Seol W, Kim JS (2003) Human zinc fingers as building blocks in the construction of artificial transcription factors. Nat Biotechnol 21:275–280

Bhatia S, Pilquil C, Roth-Albin I, Draper JS (2013) Demarcation of stable subpopulations within the pluripotent hESC compartment. PLoS One 8:e57276

Bibikova M, Golic M, Golic KG, Carroll D (2002) Targeted chromosomal cleavage and mutagenesis in Drosophila using zinc-finger nucleases. Genetics 161:1169–1175

Blancafort P, Segal DJ, Barbas CF 3rd (2004) Designing transcription factor architectures for drug discovery. Mol Pharmacol 66:1361–1371

Boch J, Scholze H, Schornack S, Landgraf A, Hahn S, Kay S, Lahaye T, Nickstadt A, Bonas U (2009) Breaking the code of DNA binding specificity of TAL-type III effectors. Science 326:1509–1512

Boissel S, Jarjour J, Astrakhan A, Adey A, Gouble A, Duchateau P, Shendure J, Stoddard B, Certo M, Baker D, Scharenberg A (2013) MegaTALs: a rare-cleaving nuclease architecture for therapeutic genome engineering. Nucleic Acids Res 42:2591–2601

Buerstedde JM, Takeda S (1991) Increased ratio of targeted to random integration after transfection of chicken B cell lines. Cell 67:179–188

Capecchi MR (2001) Generating mice with targeted mutations. Nat Med 7:1086–1090

Cermak T, Doyle EL, Christian M, Wang L, Zhang Y, Schmidt C, Baller JA, Somia NV, Bogdanove AJ, Voytas DF (2011) Efficient design and assembly of custom TALEN and other TAL effector-based constructs for DNA targeting. Nucleic Acids Res 39:e82

Chames P, Epinat J-C, Guillier S, Patin A, Lacroix E, Pâques F (2005) In vivo selection of engineered homing endonucleases using double-strand break induced homologous recombination. Nucleic Acids Res 33:e178

Chang C-J, Bouhassira EE (2012) Zinc-finger nuclease-mediated correction of α-thalassemia in iPS cells. Blood 120:3906–3914

Chapdelaine P, Pichavant C, Rousseau J, Pâques F, Tremblay JP (2010) Meganucleases can restore the reading frame of a mutated dystrophin. Gene Ther 17:846–858

Chen B, Gilbert LA, Cimini BA, Schnitzbauer J, Zhang W, Li G-W, Park J, Blackburn EH, Weissman JS, Qi LS, Huang B (2013) Dynamic imaging of genomic loci in living human cells by an optimized CRISPR/Cas system. Cell 155:1479–1491

Cheng AW, Wang H, Yang H, Shi L, Katz Y, Theunissen TW, Rangarajan S, Shivalila CS, Dadon DB, Jaenisch R (2013) Multiplexed activation of endogenous genes by CRISPR-on, an RNA-guided transcriptional activator system. Cell Res 23:1163–1171

Chevalier BS, Monnat RJ Jr, Stoddard BL (2001) The homing endonuclease I-CreI uses three metals, one of which is shared between the two active sites. Nat Struct Biol 8:312–316

Cho S-W, Kim S, Kim JM, Kim J-S (2013) Targeted genome engineering in human cells with the Cas9 RNA-guided endonuclease. Nat Biotechnol 31:230–232

Choi S, Kim Y, Shim J, Park J, Wang R, Leach S, Liu J, Deng C, Ye Z, Jang Y (2013) Efficient drug screening and gene correction for treating liver disease using patient-specific stem cells. Hepatology 57:2458–2468

Choulika A, Perrin A, Dujon B, Nicolas JF (1995) Induction of homologous recombination in mammalian chromosomes by using the I-SceI system of saccharomyces cerevisiae. Mol Cell Biol 15:1968–1973

Christian M, Cermak T, Doyle EL, Schmidt C, Zhang F, Hummel A, Bogdanove AJ, Voytas DF (2010) Targeting DNA double-strand breaks with TAL effector nucleases. Genetics 186:757–761

Coluccio A, Miselli F, Lombardo A, Marconi A, Malagoli Tagliazucchi G, Gonçalves MA, Pincelli C, Maruggi G, Del Rio M, Naldini L, Larcher F, Mavilio F, Recchia A (2013) Targeted gene addition in human epithelial stem cells by zinc-finger nuclease-mediated homologous recombination. Mol ther 21:1695–1704

Cong L, Ran FA, Cox D, Lin S, Barretto R, Habib N, Hsu PD, Wu X, Jiang W, Marraffini LA, Zhang F (2013) Multiplex genome engineering using CRISPR/Cas systems. Science 339:819–823

Cornu TI, Thibodeau-Beganny S, Guhl E, Alwin S, Eichtinger M, Joung JK, Joung JK, Cathomen T (2008) DNA-binding specificity is a major determinant of the activity and toxicity of zinc-finger nucleases. Mol Ther 16:352–358

Cradick TJ, Fine EJ, Antico CJ, Bao G (2013) CRISPR/Cas9 systems targeting β-globin and CCR5 genes have substantial off-target activity. Nucleic Acids Res 41:9584–9592

Davis RP, Ng ES, Costa M, Mossman AK, Sourris K, Elefanty AG, Stanley EG (2008) Targeting a GFP reporter gene to the MIXL1 locus of human embryonic stem cells identifies human primitive streak-like cells and enables isolation of primitive hematopoietic precursors. Blood 111:1876–1884

DeKelver R, Choi V, Moehle E, Paschon D, Hockemeyer D, Meijsing S, Sancak Y, Cui X, Steine E, Miller J, Tam P, Bartsevich V, Meng X, Rupniewski I, Gopalan S, Sun H, Pitz K, Rock J, Zhang L, Davis G, Rebar E, Cheeseman I, Yamamoto K, Sabatini D, Jaenisch R, Gregory P, Urnov F (2010) Functional genomics, proteomics, and regulatory DNA analysis in isogenic settings using zinc finger nuclease-driven transgenesis into a safe harbor locus in the human genome. Genome Res 20:1133–1142

Ding Q, Lee Y-K, Schaefer EAK, Peters DT, Veres A, Kim K, Kuperwasser N, Motola DL, Meissner TB, Hendriks WT, Trevisan M, Gupta RM, Moisan A, Banks E, Friesen M, Schinzel RT, Xia F, Tang A, Xia Y, Figueroa E, Wann A, Ahfeldt T, Daheron L, Zhang F, Rubin LL, Peng LF, Chung RT, Musunuru K, Cowan CA (2013a) A TALEN genome-editing system for generating human stem cell-based disease models. Cell Stem Cell 12:238–251

Ding Q, Regan SN, Xia Y, Oostrom LA, Cowan CA, Musunuru K (2013b) Enhanced efficiency of human pluripotent stem cell genome editing through replacing TALENs with CRISPRs. Cell Stem Cell 12:393–394

Dreier B, Segal DJ, Barbas CF 3rd (2000) Insights into the molecular recognition of the 5′-GNN-3′ family of DNA sequences by zinc finger domains. J Mol Biol 303:489–502

Dreier B, Beerli RR, Segal DJ, Flippin JD, Barbas CF 3rd (2001) Development of zinc finger domains for recognition of the 5′-ANN-3′ family of DNA sequences and their use in the construction of artificial transcription factors. J Biol Chem 276:29466–29478

Ebina H, Misawa N, Kanemura Y, Koyanagi Y (2013) Harnessing the CRISPR/Cas9 system to disrupt latent HIV-1 provirus. Sci Rep 3:2510

Epinat J-C, Arnould S, Chames P, Rochaix P, Desfontaines D, Puzin C, Patin A, Zanghellini A, Pâques F, Lacroix E (2003) A novel engineered meganuclease induces homologous recombination in yeast and mammalian cells. Nucleic Acids Res 31:2952–2962

Evans MJ, Kaufman MH (1981) Establishment in culture of pluripotential cells from mouse embryos. Nature 292:154–156

Folger KR, Wong EA, Wahl G, Capecchi MR (1982) Patterns of integration of DNA microinjected into cultured mammalian cells: evidence for homologous recombination between injected plasmid DNA molecules. Mol Cell Biol 2:1372–1387

Fu Y, Sander JD, Reyon D, Cascio VM, Joung JK (2014) Improving CRISPR-Cas nuclease specificity using truncated guide RNAs. Nat Biotechnol 32:279–284

Fujita T, Fujii H (2013) Efficient isolation of specific genomic regions and identification of associated proteins by engineered DNA-binding molecule-mediated chromatin immunoprecipitation (enChIP) using CRISPR. Biochem Biophys Res Commun 439:132–136

Fujita T, Asano Y, Ohtsuka J, Takada Y, Saito K, Ohki R, Fujii H (2013) Identification of telomere-associated molecules by engineered DNA-binding molecule-mediated chromatin immunoprecipitation (enChIP). Sci Rep 3:3171

Garçon L, Ge J, Manjunath SH, Mills JA, Apicella M, Parikh S, Sullivan LM, Podsakoff GM, Gadue P, French DL, Mason PJ, Bessler M, Weiss MJ (2013) Ribosomal and hematopoietic defects in induced pluripotent stem cells derived from diamond Blackfan anemia patients. Blood 122:912–921

Genovese P, Schiroli G, Escobar G, Di Tomaso T, Firrito C, Calabria A, Moi D, Mazzieri R, Bonini C, Holmes MC, Gregory PD, van der Burg M, Gentner B, Montini E, Lombardo A, Naldini L (2014) Targeted genome editing in human repopulating haematopoietic stem cells. Nature 510:235–240

Gilbert LA, Larson MH, Morsut L, Liu Z, Brar GA, Torres SE, Stern-Ginossar N, Brandman O, Whitehead EH, Doudna JA, Lim WA, Weissman JS, Qi LS (2013) CRISPR-mediated modular RNA-guided regulation of transcription in eukaryotes. Cell 154:442–451

Gonzalez F, Zhu Z, Shi ZD, Lelli K, Verma N, Li QV, Huangfu D (2014) An iCRISPR platform for rapid, multiplexable, and inducible genome editing in human pluripotent stem cells. Cell Stem Cell 15:1–12

Grizot S, Smith J, Daboussi F, Prieto J, Redondo P, Merino N, Villate M, Thomas S, Lemaire L, Montoya G, Blanco FJ, Pâques F, Duchateau P (2009) Efficient targeting of a SCID gene by an engineered single-chain homing endonuclease. Nucleic Acids Res 37

Grizot S, Epinat J-C, Thomas S, Duclert A, Rolland S, Pâques F, Duchateau P (2010) Generation of redesigned homing endonucleases comprising DNA-binding domains derived from two different scaffolds. Nucleic Acids Res 38:2006–2018

Hockemeyer D, Soldner F, Beard C, Gao Q, Mitalipova M, DeKelver RC, Katibah GE, Amora R, Boydston EA, Zeitler B, Meng X, Miller JC, Zhang L, Rebar EJ, Gregory PD, Urnov FD, Jaenisch R (2009) Efficient targeting of expressed and silent genes in human ESCs and iPSCs using zinc-finger nucleases. Nat Biotechnol 27:851–857

Hockemeyer D, Wang H, Kiani S, Lai CS, Gao Q, Cassady JP, Cost GJ, Zhang L, Santiago Y, Miller JC, Zeitler B, Cherone JM, Meng X, Hinkley SJ, Rebar EJ, Gregory PD, Urnov FD, Jaenisch R (2011) Genetic engineering of human pluripotent cells using TALE nucleases. Nat Biotechnol 29:731–734

Holkers M, Maggio I, Liu J, Janssen JM, Miselli F, Mussolino C, Recchia A, Cathomen T, Gonçalves MAFV (2013) Differential integrity of TALE nuclease genes following adenoviral and lentiviral vector gene transfer into human cells. Nucleic Acids Res 41:e63

Holt N, Wang J, Kim K, Friedman G, Wang X, Taupin V, Crooks GM, Kohn DB, Gregory PD, Holmes MC, Cannon PM (2010) Human hematopoietic stem/progenitor cells modified by zinc-finger nucleases targeted to CCR5 control HIV-1 in vivo. Nat Biotechnol 28:839–847

Hotta A, Cheung AY, Farra N, Vijayaragavan K, Seguin CA, Draper JS, Pasceri P, Maksakova IA, Mager DL, Rossant J, Bhatia M, Ellis J (2009) Isolation of human iPS cells using EOS lentiviral vectors to select for pluripotency. Nat Methods 6:370–376

Hsu PD, Scott DA, Weinstein JA, Ran FA, Konermann S, Agarwala V, Li Y, Fine EJ, Wu X, Shalem O, Cradick TJ, Marraffini LA, Bao G, Zhang F (2013) DNA targeting specificity of RNA-guided Cas9 nucleases. Nat Biotechnol 31:827–832

Isalan M, Klug A, Choo Y (1998) Comprehensive DNA recognition through concerted interactions from adjacent zinc fingers. Biochemistry 37:12026–12033

Jaenisch R (1977) Germ line integration of moloney leukemia virus: effect of homozygosity at the m-mulV locus. Cell 12:691–696

Jia J, Jin Y, Bian T, Wu D, Yang L, Terada N, Wu W, Jin S (2014) Bacterial delivery of TALEN proteins for human genome editing. PLoS One 9:e91547

Jiang J, Jing Y, Cost G, Chiang J, Kolpa H, Cotton A, Carone D, Carone B, Shivak D, Guschin D, Pearl J, Rebar E, Byron M, Gregory P, Brown C, Urnov F, Hall L, Lawrence J (2013a) Translating dosage compensation to trisomy 21. Nature 500:296–300

Jiang W, Bikard D, Cox D, Zhang F, Marraffini LA (2013b) RNA-guided editing of bacterial genomes using CRISPR-Cas systems. Nat Biotechnol 31:233–239

Jinek M, Chylinski K, Fonfara I, Hauer M, Doudna JA, Charpentier E (2012) A programmable dual-RNA-guided DNA endonuclease in adaptive bacterial immunity. Science 337:816–821

Jinek M, East A, Cheng A, Lin S, Ma E, Doudna J (2013) RNA-programmed genome editing in human cells. Elife 2:e00471

Kim JA, Cho K, Shin MS, Lee WG, Jung N, Chung C, Chang JK (2008) A novel electroporation method using a capillary and wire-type electrode. Biosens Bioelectron 23:1353–1360

Kim HJ, Lee HJ, Kim H, Cho SW, Kim JS (2009) Targeted genome editing in human cells with zinc finger nucleases constructed via modular assembly. Genome Res 19:1279–1288

Kim Y, Kweon J, Kim A, Chon JK, Yoo JY, Kim HJ, Kim S, Lee C, Jeong E, Chung E, Kim D, Lee MS, Go EM, Song HJ, Kim H, Cho N, Bang D, Kim S, Kim J-S (2013a) A library of TAL effector nucleases spanning the human genome. Nat Biotechnol 31:251–258

Kim Y-K, Wee G, Park J, Kim J, Baek D, Kim J-S, Kim VN (2013b) TALEN-based knockout library for human microRNAs. Nat Struct Mol Biol 20:1458–1464

Kim S, Kim D, Cho S-W, Kim J, Kim J-S (2014) Highly efficient RNA-guided genome editing in human cells via delivery of purified Cas9 ribonucleoproteins. Genome Res 24:1012–1019

Koike-Yusa H, Li Y, Tan E-P, Velasco-Herrera MDC, Yusa K (2014) Genome-wide recessive genetic screening in mammalian cells with a lentiviral CRISPR-guide RNA library. Nat Biotechnol 32:267–273

Konermann S, Brigham MD, Trevino AE, Hsu PD, Heidenreich M, Cong L, Platt RJ, Scott DA, Church GM, Zhang F (2013) Optical control of mammalian endogenous transcription and epigenetic states. Nature 500:472–476

Lee HJ, Kweon J, Kim E, Kim S, Kim J-S (2012) Targeted chromosomal duplications and inversions in the human genome using zinc finger nucleases. Genome Res 22:539–548

Li H, Haurigot V, Doyon Y, Li T, Wong SY, Bhagwat AS, Malani N, Anguela XM, Sharma R, Ivanciu L, Murphy SL, Finn JD, Khazi FR, Zhou S, Paschon DE, Rebar EJ, Bushman FD, Gregory PD, Holmes MC, High KA (2011) In vivo genome editing restores haemostasis in a mouse model of haemophilia. Nature 475:217–221

Li S-J, Shi R-Z, Bai Y-P, Hong D, Yang W, Wang X, Mo L, Zhang G-G (2013) Targeted introduction of tissue plasminogen activator (TPA) at the AAVS1 locus in mesenchymal stem cells (MSCs) and its stable and effective expression. Biochem Biophys Res Commun 437:74–78

Lin Y, Cradick TJ, Brown MT, Deshmukh H, Ranjan P, Sarode N, Wile BM, Vertino PM, Stewart FJ, Bao G (2014) CRISPR/Cas9 systems have off-target activity with insertions or deletions between target DNA and guide RNA sequences. Nucleic Acids Res 42:7473–7485

Lisa Li H, Nakano T, Hotta A (2014) Genetic correction using engineered nucleases for gene therapy applications. Dev Growth Differ 56:63–77

Lombardo A, Genovese P, Beausejour CM, Colleoni S, Lee Y-L, Kim KA, Ando D, Urnov FD, Galli C, Gregory PD, Holmes MC, Naldini L (2007) Gene editing in human stem cells using zinc finger nucleases and integrase-defective lentiviral vector delivery. Nat Biotechnol 25:1298–1306

Lombardo A, Cesana D, Genovese P, Di Stefano B, Provasi E, Colombo DF, Neri M, Magnani Z, Cantore A, Lo Riso P, Damo M, Pello OM, Holmes MC, Gregory PD, Gritti A, Broccoli V, Bonini C, Naldini L (2011) Site-specific integration and tailoring of cassette design for sustainable gene transfer. Nat Methods 8:861–869

Ma H, Reyes-Gutierrez P, Pederson T (2013) Visualization of repetitive DNA sequences in human chromosomes with transcription activator-like effectors. Proc Natl Acad Sci U S A 110:21048–21053

Ma N, Liao B, Zhang H, Wang L, Shan Y, Xue Y, Huang K, Chen S, Zhou X, Chen Y, Pei D, Pan G (2014) Transcription activator-like effector nuclease (TALEN)-mediated gene correction in integration-free β-thalassemia induced pluripotent stem cells. J Biol Chem 288:34671–34679

Mae S, Shono A, Shiota F, Yasuno T, Kajiwara M, Gotoda-Nishimura N, Arai S, Sato-Otubo A, Toyoda T, Takahashi K, Nakayama N, Cowan CA, Aoi T, Ogawa S, McMahon AP, Yamanaka S, Osafune K (2013) Monitoring and robust induction of nephrogenic intermediate mesoderm from human pluripotent stem cells. Nat Commun 4:1367

Maeder ML, Thibodeau-Beganny S, Osiak A, Wright DA, Anthony RM, Eichtinger M, Jiang T, Foley JE, Winfrey RJ, Townsend JA, Unger-Wallace E, Sander JD, Muller-Lerch F, Fu F, Pearlberg J, Gobel C, Dassie JP, Pruett-Miller SM, Porteus MH, Sgroi DC, Iafrate AJ, Dobbs D, McCray PB Jr, Cathomen T, Voytas DF, Joung JK (2008) Rapid "open-source" engineering of customized zinc-finger nucleases for highly efficient gene modification. Mol Cell 31:294–301

Maeder ML, Linder SJ, Cascio VM, Fu Y, Ho QH, Joung JK (2013a) CRISPR RNA-guided activation of endogenous human genes. Nat Methods 10:977–979

Maeder ML, Linder SJ, Reyon D, Angstman JF, Fu Y, Sander JD, Joung JK (2013b) Robust, synergistic regulation of human gene expression using TALE activators. Nat Methods 10(3): 243–245

Maggio I, Holkers M, Liu J, Janssen JM, Chen X, Gonçalves MAFV (2014) Adenoviral vector delivery of RNA-guided CRISPR/Cas9 nuclease complexes induces targeted mutagenesis in a diverse array of human cells. Sci Rep 4:5105

Maier DA, Brennan AL, Jiang S, Binder-Scholl GK, Lee G, Plesa G, Zheng Z, Cotte J, Carpenito C, Wood T, Spratt SK, Ando D, Gregory P, Holmes MC, Perez EE, Riley JL, Carroll RG, June CH, Levine BL (2013) Efficient clinical scale gene modification via zinc finger nuclease-targeted disruption of the HIV co-receptor CCR5. Hum Gene Ther 24:245–258

Mali P, Aach J, Stranges PB, Esvelt KM, Moosburner M, Kosuri S, Yang L, Church GM (2013a) CAS9 transcriptional activators for target specificity screening and paired nickases for cooperative genome engineering. Nat Biotechnol 31:833–838

Mali P, Yang L, Esvelt KM, Aach J, Guell M, DiCarlo JE, Norville JE, Church GM (2013b) RNA-guided human genome engineering via Cas9. Science 339:823–826

Mendenhall EM, Williamson KE, Reyon D, Zou JY, Ram O, Joung JK, Bernstein BE (2013) Locus-specific editing of histone modifications at endogenous enhancers. Nat Biotechnol 31:1133–1136

Miller JC, Tan S, Qiao G, Barlow KA, Wang J, Xia DF, Meng X, Paschon DE, Leung E, Hinkley SJ, Dulay GP, Hua KL, Ankoudinova I, Cost GJ, Urnov FD, Zhang HS, Holmes MC, Zhang L, Gregory PD, Rebar EJ (2011) A TALE nuclease architecture for efficient genome editing. Nat Biotechnol 29:143–148

Miyanari Y, Ziegler-Birling C, Torres-Padilla ME (2013) Live visualization of chromatin dynamics with fluorescent TALEs. Nat Struct Mol Biol 20:1321–1324

Moscou MJ, Bogdanove AJ (2009) A simple cipher governs DNA recognition by TAL effectors. Science 326:1501

Mussolino C, Morbitzer R, Lütge F, Dannemann N, Lahaye T, Cathomen T (2011) A novel TALE nuclease scaffold enables high genome editing activity in combination with low toxicity. Nucleic Acids Res 39:9283–9293

Neumann E, Schaefer-Ridder M, Wang Y, Hofschneider PH (1982) Gene transfer into mouse lyoma cells by electroporation in high electric fields. EMBO J 1:841–845

Osborn MJ, Starker CG, McElroy AN, Webber BR, Riddle MJ, Xia L, DeFeo AP, Gabriel R, Schmidt M, von Kalle C, Carlson DF, Maeder ML, Joung JK, Wagner JE, Voytas DF, Blazar BR, Tolar J (2013) TALEN-based gene correction for epidermolysis bullosa. Mol Ther 21:1151–1159

Ousterout DG, Perez-Pinera P, Thakore PI, Kabadi AM, Brown MT, Qin X, Fedrigo O, Mouly V, Tremblay JP, Gersbach CA (2013) Reading frame correction by targeted genome editing restores dystrophin expression in cells from Duchenne muscular dystrophy patients. Mol Ther 21:1718–1726

Pabo CO, Peisach E, Grant RA (2001) Design and selection of novel Cys2His2 zinc finger proteins. Annu Rev Biochem 70:313–340

Park CY, Kim J, Kweon J, Son JS, Lee JS, Yoo JE, Cho SR, Kim JH, Kim JS, Kim DW (2014) Targeted inversion and reversion of the blood coagulation factor 8 gene in human iPS cells using TALENs. Proc Natl Acad Sci U S A 111:9253–9258

Perez EE, Wang J, Miller JC, Jouvenot Y, Kim KA, Liu O, Wang N, Lee G, Bartsevich VV, Lee Y-L, Guschin DY, Rupniewski I, Waite AJ, Carpenito C, Carroll RG, Orange JS, Urnov FD, Rebar EJ, Ando D, Gregory PD, Riley JL, Holmes MC, June CH (2008) Establishment of HIV-1 resistance in CD4+ T cells by genome editing using zinc-finger nucleases. Nat Biotechnol 26:808–816

Perez-Pinera P, Kocak DD, Vockley CM, Adler AF, Kabadi AM, Polstein LR, Thakore PI, Glass KA, Ousterout DG, Leong KW, Guilak F, Crawford GE, Reddy TE, Gersbach CA (2013) RNA-guided gene activation by CRISPR-Cas9-based transcription factors. Nat Methods 10:973–976

Phang R-Z, Tay FC, Goh S-L, Lau C-H, Zhu H, Tan W-K, Liang Q, Chen C, Du S, Li Z, Tay JC-K, Wu C, Zeng J, Fan W, Toh HC, Wang S (2013) Zinc finger nuclease-expressing baculoviral vectors mediate targeted genome integration of reprogramming factor genes to facilitate the generation of human induced pluripotent stem cells. Stem Cells Transl Med 2:935–945

Provasi E, Genovese P, Lombardo A, Magnani Z, Liu P, Reik A, Chu V, Paschon D, Zhang L, Kuball J, Camisa B, Bondanza A, Casorati G, Ponzoni M, Ciceri F, Bordignon C, Greenberg P, Holmes M, Gregory P, Naldini L, Bonini C (2012) Editing T cell specificity towards leukemia by zinc finger nucleases and lentiviral gene transfer. Nat Med 18:807–815

Qi LS, Larson MH, Gilbert LA, Doudna JA, Weissman JS, Arkin AP, Lim WA (2013) Repurposing CRISPR as an RNA-guided platform for sequence-specific control of gene expression. Cell 152:1173–1183

Qian K, Huang CL, Chen H, Blackbourn LW 4th, Chen Y, Cao J, Yao L, Sauvey C, Du Z, Zhang SC (2014) A simple and efficient system for regulating gene expression in human pluripotent stem cells and derivatives. Stem Cells 32:1230–1238

Ramakrishna S, Kwaku Dad A-B, Beloor J, Gopalappa R, Lee S-K, Kim H (2014) Gene disruption by cell-penetrating peptide-mediated delivery of Cas9 protein and guide RNA. Genome Res 24:1020–1027

Ran FA, Hsu PD, Lin C-Y, Gootenberg JS, Konermann S, Trevino AE, Scott DA, Inoue A, Matoba S, Zhang Y, Zhang F (2013) Double nicking by RNA-guided CRISPR Cas9 for enhanced genome editing specificity. Cell 154:1380–1389

Reyon D, Tsai SQ, Khayter C, Foden JA, Sander JD, Joung JK (2012) FLASH assembly of TALENs for high-throughput genome editing. Nat Biotechnol 30:460–465

Río P, Baños R, Lombardo A, Quintana-Bustamante O, Alvarez L, Garate Z, Genovese P, Almarza E, Valeri A, Díez B, Navarro S, Torres Y, Trujillo JP, Murillas R, Segovia JC, Samper E, Surrallés J, Gregory PD, Holmes MC, Naldini L, Bueren JA (2014) Targeted gene therapy and cell reprogramming in Fanconi anemia. EMBO Mol Med 6:835–848

Rouet P, Smih F, Jasin M (1994a) Expression of a site-specific endonuclease stimulates homologous recombination in mammalian cells. Proc Natl Acad Sci U S A 91:6064–6068

Rouet P, Smih F, Jasin M (1994b) Introduction of double-strand breaks into the genome of mouse cells by expression of a rare-cutting endonuclease. Mol Cell Biol 14:8096–8106

Rousseau J, Chapdelaine P, Boisvert S, Almeida LP, Corbeil J, Montpetit A, Tremblay JP (2011) Endonucleases: tools to correct the dystrophin gene. J Gene Med 13:522–537

Ruby KM, Zheng B (2009) Gene targeting in a HUES line of human embryonic stem cells via electroporation. Stem Cells 27:1496–1506

Rusk N (2014) CRISPRs and epigenome editing. Nat Methods 11:28

Sadowski I, Ma J, Triezenberg S, Ptashne M (1988) GAL4-VP16 is an unusually potent transcriptional activator. Nature 335:563–564

Sakuma T, Hosoi S, Woltjen K, Suzuki K-I, Kashiwagi K, Wada H, Ochiai H, Miyamoto T, Kawai N, Sasakura Y, Matsuura S, Okada Y, Kawahara A, Hayashi S, Yamamoto T (2013) Efficient TALEN construction and evaluation methods for human cell and animal applications. Genes Cells 18:315–326

Sakuma T, Nishikawa A, Kume S, Chayama K, Yamamoto T (2014) Multiplex genome engineering in human cells using all-in-one CRISPR/Cas9 vector system. Sci Rep 4:5400

Scherer WF, Syverton JT, Gey GO (1953) Studies on the propagation in vitro of poliomyelitis viruses. IV. Viral multiplication in a stable strain of human malignant epithelial cells (strain HeLa) derived from an epidermoid carcinoma of the cervix. J Exp Med 97:695–710

Schwank G, Koo B-K, Sasselli V, Dekkers JF, Heo I, Demircan T, Sasaki N, Boymans S, Cuppen E, van der Ent CK, Nieuwenhuis EES, Beekman JM, Clevers H (2013) Functional repair of CFTR by CRISPR/Cas9 in intestinal stem cell organoids of cystic fibrosis patients. Cell Stem Cell 13:653–658

Sebastiano V, Maeder ML, Angstman JF, Haddad B, Khayter C, Yeo DT, Goodwin MJ, Hawkins JS, Ramirez CL, Batista LFZ, Artandi SE, Wernig M, Joung JK (2011) In situ genetic correction of the sickle cell anemia mutation in human induced pluripotent stem cells using engineered zinc finger nucleases. Stem Cells 29:1717–1726

Segal DJ, Dreier B, Beerli RR, Barbas CF 3rd (1999) Toward controlling gene expression at will: selection and design of zinc finger domains recognizing each of the 5′-GNN-3′ DNA target sequences. Proc Natl Acad Sci U S A 96:2758–2763

Shalem O, Sanjana NE, Hartenian E, Shi X, Scott DA, Mikkelsen TS, Heckl D, Ebert BL, Root DE, Doench JG, Zhang F (2014) Genome-scale CRISPR-Cas9 knockout screening in human cells. Science 343:84–87

Shen B, Zhang W, Zhang J, Zhou J, Wang J, Chen L, Wang L, Hodgkins A, Iyer V, Huang X, Skarnes WC (2014) Efficient genome modification by CRISPR-Cas9 nickase with minimal off-target effects. Nat Methods 11:399–402

Silva GH, Dalgaard JZ, Belfort M, Van Roey P (1999) Crystal structure of the thermostable archaeal intron-encoded endonuclease I-DmoI. J Mol Biol 286:1123–1136

Smith J, Bibikova M, Whitby FG, Reddy AR, Chandrasegaran S, Carroll D (2000) Requirements for double-strand cleavage by chimeric restriction enzymes with zinc finger DNA-recognition domains. Nucleic Acids Res 28:3361–3369

Smithies O, Gregg RG, Boggs SS, Koralewski MA, Kucherlapati RS (1985) Insertion of DNA sequences into the human chromosomal beta-globin locus by homologous recombination. Nature 317:230–234

Soldner F, Laganière J, Cheng AW, Hockemeyer D, Gao Q, Alagappan R, Khurana V, Golbe LI, Myers RH, Lindquist S, Zhang L, Guschin D, Fong LK, Vu BJ, Meng X, Urnov FD, Rebar EJ, Gregory PD, Zhang HS, Jaenisch R (2011) Generation of isogenic pluripotent stem cells differing exclusively at two early onset Parkinson point mutations. Cell 146:318–331

Song H, Chung S-K, Xu Y (2010) Modeling disease in human ESCs using an efficient BAC-based homologous recombination system. Cell Stem Cell 6:80–89

Takahashi K, Yamanaka S (2006) Induction of pluripotent stem cells from mouse embryonic and adult fibroblast cultures by defined factors. Cell 126:663–676

Takahashi K, Tanabe K, Ohnuki M, Narita M, Ichisaka T, Tomoda K, Yamanaka S (2007) Induction of pluripotent stem cells from adult human fibroblasts by defined factors. Cell 131:861–872

Tay FC, Tan W-K, Goh S-L, Ramachandra CJA, Lau C-H, Zhu H, Chen C, Du S, Phang R-Z, Shahbazi M, Fan W, Wang S (2013) Targeted transgene insertion into the AAVS1 locus driven by baculoviral vector-mediated zinc finger nuclease expression in human-induced pluripotent stem cells. J Gene Med 15:384–395

Tebas P, Stein D, Tang WW, Frank I, Wang SQ, Lee G, Spratt SK, Surosky RT, Giedlin MA, Nichol G, Holmes MC, Gregory PD, Ando DG, Kalos M, Collman RG, Binder-Scholl G, Plesa G, Hwang WT, Levine BL, June CH (2014) Gene editing of CCR5 in autologous CD4 T cells of persons infected with HIV. N Engl J Med 370:901–910

Thomas KR, Capecchi MR (1987) Site-directed mutagenesis by gene targeting in mouse embryo-derived stem cells. Cell 51:503–512

Thomson JA, Itskovitz-Eldor J, Shapiro SS, Waknitz MA, Swiergiel JJ, Marshall VS, Jones JM (1998) Embryonic stem cell lines derived from human blastocysts. Science 282:1145–1147

Torikai H, Reik A, Liu P, Zhou Y, Zhang L, Maiti S, Huls H, Miller J, Kebriaei P, Rabinovitch B, Lee D, Champlin R, Bonini C, Naldini L, Rebar E, Gregory P, Holmes M, Cooper L (2012) A foundation for universal T-cell based immunotherapy: T cells engineered to express a CD19-specific chimeric-antigen-receptor and eliminate expression of endogenous TCR. Blood 119:5697–5705

Tsai SQ, Wyvekens N, Khayter C, Foden JA, Thapar V, Reyon D, Goodwin MJ, Aryee MJ, Joung JK (2014) Dimeric CRISPR RNA-guided FokI nucleases for highly specific genome editing. Nat Biotechnol 32:569–576

Urbach A, Schuldiner M, Benvenisty N (2004) Modeling for Lesch-Nyhan disease by gene targeting in human embryonic stem cells. Stem Cells 22:635–641

Urnov FD, Miller JC, Lee Y-L, Beausejour CM, Rock JM, Augustus S, Jamieson AC, Porteus MH, Gregory PD, Holmes MC (2005) Highly efficient endogenous human gene correction using designed zinc-finger nucleases. Nature 435:646–651

Voigt P, Reinberg D (2013) Epigenome editing. Nat Biotechnol 31:1097–1099

Wang Y, Zhang WY, Hu S, Lan F, Lee AS, Huber B, Lisowski L, Liang P, Huang M, de Almeida PE, Won JH, Sun N, Robbins RC, Kay MA, Urnov FD, Wu JC (2012) Genome editing of human embryonic stem cells and induced pluripotent stem cells with zinc finger nucleases for cellular imaging. Circ Res 111:1494–1503

Wang T, Wei JJ, Sabatini DM, Lander ES (2014) Genetic screens in human cells using the CRISPR-Cas9 system. Science 343:80–84

Yao Y, Nashun B, Zhou T, Qin L, Qin L, Zhao S, Xu J, Esteban M, Chen X (2011) Generation of CD34+ cells from CCR5-disrupted human embryonic and induced pluripotent stem cells. Hum Gene Ther 23:238–242

Ye L, Wang J, Beyer AI, Teque F, Cradick TJ, Qi Z, Chang JC, Bao G, Muench MO, Yu J, Levy JA, Kan YW (2014) Seamless modification of wild-type induced pluripotent stem cells to the natural CCR5Delta32 mutation confers resistance to HIV infection. Proc Natl Acad Sci U S A 111:9591–9596

Yusa K, Rashid ST, Strick-Marchand H, Varela I, Liu P-Q, Paschon DE, Miranda E, Ordóñez A, Hannan NRF, Rouhani FJ, Darche S, Alexander G, Marciniak SJ, Fusaki N, Hasegawa M, Holmes MC, Di Santo JP, Lomas DA, Bradley A, Vallier L (2011) Targeted gene correction of α1-antitrypsin deficiency in induced pluripotent stem cells. Nature 478:391–394

Zhou Y, Zhu S, Cai C, Yuan P, Li C, Huang Y, Wei W (2014) High-throughput screening of a CRISPR/Cas9 library for functional genomics in human cells. Nature 509:487–491

Zhu H, Lau C-H, Goh S-L, Liang Q, Chen C, Du S, Phang R-Z, Tay FC, Tan W-K, Li Z, Tay JC-K, Fan W, Wang S (2013) Baculoviral transduction facilitates TALEN-mediated targeted transgene integration and Cre/LoxP cassette exchange in human-induced pluripotent stem cells. Nucleic Acids Res 41:e180

Zou J, Maeder ML, Mali P, Pruett-Miller SM, Thibodeau-Beganny S, Chou B-K, Chen G, Ye Z, Park I-H, Daley GQ, Porteus MH, Joung JK, Cheng L (2009) Gene targeting of a disease-related gene in human induced pluripotent stem and embryonic stem cells. Cell Stem Cell 5:97–110

Zou J, Mali P, Huang X, Dowey SN, Cheng L (2011a) Site-specific gene correction of a point mutation in human iPS cells derived from an adult patient with sickle cell disease. Blood 118: 4599–4608

Zou J, Sweeney CL, Chou BK, Choi U, Pan J, Wang H, Dowey SN, Cheng L, Malech HL (2011b) Oxidase-deficient neutrophils from X-linked chronic granulomatous disease iPS cells: functional correction by zinc finger nuclease-mediated safe harbor targeting. Blood 117:5561–5572

Zwaka TP, Thomson JA (2003) Homologous recombination in human embryonic stem cells. Nat Biotechnol 21:319–321

Chapter 4
Genome Editing in Nematode

Takuma Sugi

Abstract Nematode species, such as *Caenorhabditis elegans* (*C. elegans*), *Caenorhabditis briggsae* (*C. briggsae*), *Pristionchus pacificus* (*P. pacificus*), are a valuable and widely used model animal owing to their rapid growth, ease of genetic handling and transparency that facilitates microscopic observation of fluorescent proteins. However, previous conventional methods to generate mutations in the genome of nematodes, including chemical mutagenesis and imprecise excision of transposons, all had relied on recovering mutations in large-scale mutagenesis screens. The recent development of genome engineering in nematode species overcomes the prior laborious and time-consuming procedures and promises to dramatically advance our ability to create locus-specific mutations at will. In this chapter, I highlight the recently developed genome editing technique using TALEN and CRISPR/Cas9 systems in nematodes and their future perspectives.

Keywords *C. elegans* • Cell-specific promoter • Nematode

4.1 Introduction

1 mm, free-living nematode *C. elegans* was introduced by Sydney Brenner in 1963 as a powerful model organism (Brenner 1974) (Fig. 4.1). Scientists have used this tiny creature to uncover functions of genes in development and cellular biology for more than 50 years. One of characteristic features which made nematode an attractive experimental model for scientists is its simple morphology; the adult hermaphrodite has only 959 cells. The lineage of the 959 somatic cells was traced through its transparent cuticle, allowing for determination of cell fate (Sulston et al. 1983). This future has remarkably contributed to researches of developmental processes. Of 959 cells, 302 are neurons and constitute a compact neural circuit. Their wiring

T. Sugi (✉)
Institute for Integrated Cell-Material Sciences (WPI-iCeMS), Kyoto University, Yoshida-Honmachi, Sakyo-ku, Kyoto 606-8501, Japan

PRESTO, Japanese Science and Technology Agency, 4-1-8 Honcho, Kawaguchi, Saitama 332-0012, Japan
e-mail: tsugi@icems.kyoto-u.ac.jp

T. Yamamoto (ed.), *Targeted Genome Editing Using Site-Specific Nucleases*, DOI 10.1007/978-4-431-55227-7_4

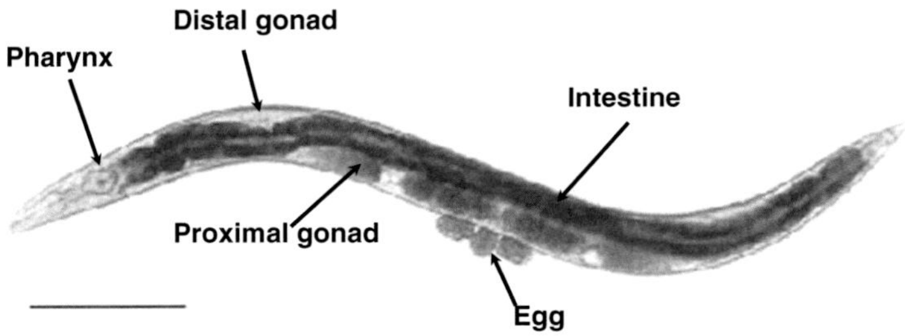

Fig. 4.1 Anatomy of nematode *C. elegans*. DIC image of an adult hermaphrodite of *C. elegans*. ZFN, TALEN, or CRSIPR/Cas9 are injected into the distal gonad of worms. Scale bar, 200 μm

and connectivity have been mapped completely by the electron microscopy (White et al. 1986). This nervous system is well-adapted to respond to a variety of sensory modalities, including chemosensation, mechanosensation, and thermosensation. In addition, nematodes can alter a behavioral strategy based on previously experienced environmental conditions. Therefore, the nervous system of nematodes, despite its remarkable simplicity, serves as interrogation of neural circuitry underlying behavioral plasticity. In addition to these futures, the small size, short life cycle (less than three days) and ability to survive being frozen for long-term storage in nematode make it highly amenable to laboratory research. Thus, nematode is an ideal model organism that provides opportunity to investigate various biological processes conserved in other model animals such as mammals.

Nematode, especially *C. elegans*, is a genetically tractable model system, in which many genetic resources and tools have been developed for this organism. As of 2012, among the 20,377 predicted protein-coding genes in *C. elegans*, at least approximately 38 % of which have homologues in humans (Shaye and Greenwald 2011), 6,764 genes have either deletions or null mutations, and these mutants are available for genetic experiments (*C. elegans* Deletion Mutant Consortium 2012). Furthermore, numerous methods for modifying the genome have been developed. These include the random introduction of mutations through chemical means, and the random insertion of transgenes in the genome by integration of extrachromosomal arrays, or as low copy insertions through microparticle bombardment (Kage-Nakadai et al. 2012). More recently, methods based on mobilization of the *Mos1* transposon have enabled targeted genome engineering (Frøkjaer-Jensen et al. 2008; Robert and Bessereau 2007). However, only a limited fraction of the genome is accessible for *Mos1*-based genome engineering, as the efficiency of homologous recombination declines with increased distance from the double strand break site. The ~13,300 *Mos1* insertions currently available to the community target ~40 % of all *C. elegans* genes, and in many cases only a limited region of the gene can be targeted.

Targeted genome editing using zinc finger nucleases (ZFNs), transcription activator-like effector (TALE) nucleases (TALENs), and RNA-guided clustered regulatory interspaced short palindromic repeats (CRISPR)/Cas9 endonuclease systems has been adapted to nematode species as a powerful method for creating locus-specific mutations (Frøkjaer-Jensen 2013). In this chapter, I review recent implementations of genome editing approaches in nematode species.

4.2 Genome Engineering in Nematode Species Using ZFNs, TALENs, and CRISPR/Cas9 System

4.2.1 Targeted Knockout by ZFNs and Mos1 *Transposon*

The genome engineering in *C. elegans* was initiated with the pioneer work by Jorgensen and his colleagues (Morton et al. 2006). They initially designed a synthetic target for a previously characterized zinc-finger nuclease. The synthetic target and plasmid that expresses the nuclease under the control of heat-shock promoter were coinjected into the germ line to generate an extrachromosomal, mutlicopy transgene array. The more than 20 % of the target sites were mutated after induction of the nuclease by heat-shock treatment. Then, an endogenous genomic sequence was targeted with a pair of specifically designed nucleases. In this case, approximately 20 % of the target site was also mutated after the nuclease induction. This study showed for the first time that custom-engineered ZFNs were capable of generating targeted double-strand breaks in nematode DNA at high frequency. Although repair of these breaks by non-homologous end joining (NHEJ) often generated mutations at the targeted locus, the procedure shown in their study had relatively little practical value as a genetic tool, because mutations were generated in somatic cells and not inherited to offspring. Therefore, this paper suggested that direct injection of mRNA to germ-line would be effective to obtain an inherited mutant.

The *Drosophila* element *Mos1* was used as an alternative tool to engineer the target genome sequence (Robert and Bessereau 2007). First, a transgene containing sequences homologous to the broken chromosomal region was designed as a repair template. The repair template was then injected into the germ-line of *Mos1*-inserted mutant together with a vector enabling the expression of the *Mos* transposase under the control of the heat-shock promoter *hsp-16.48*. The double-strand break repair could be achieved by transgene-instructed gene conversion after heat-shock treatment, and mutations engineered in the transgene could be copied to specific locus at high frequency. The mutation was successfully inherited to offspring. The prerequisite of this technique is to have a *Mos1* element inserted into the target genomic region to be engineered. Currently, a European effort (http://elegans.imbb.forth.gr/nemagenetag/) is in progress to generate a comprehensive *Mos1* insertion mutant library. Despite these progresses, the ability to modify the genome had been limited to locations near endogenous or exogenous transposons.

4.2.2 Targeted Knockout Using TALEN and CRISPR/Cas9 System

In 2011, the revolutionized genome editing technology using TALEN was applied to nematode species for the first time (Wood et al. 2011). TALENs comprise a non-specific FokI nuclease domain that is fused to a customizable TALE domain, which recognizes a predictable DNA sequence (Mussolino and Cathomen 2012).

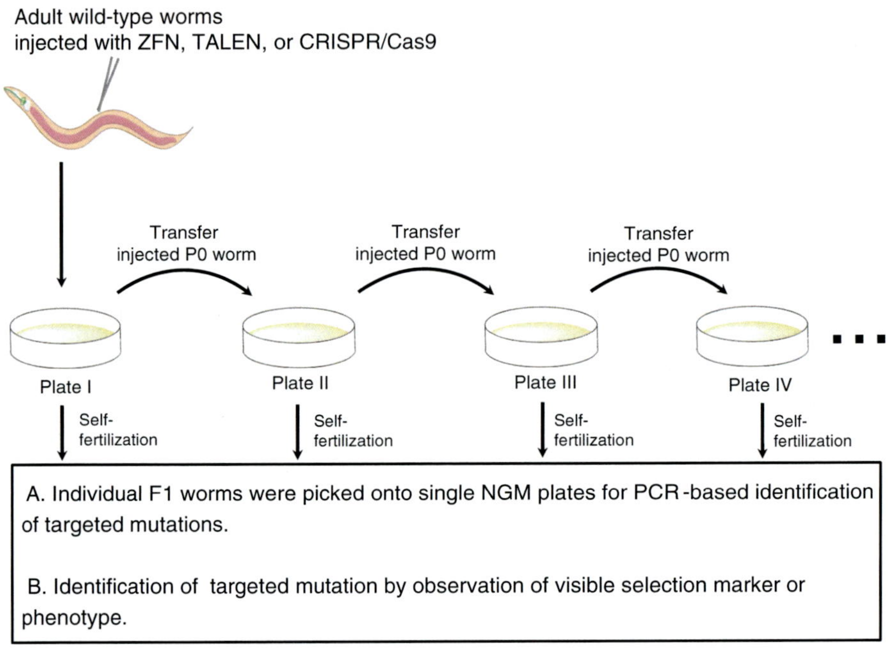

Fig. 4.2 An experimental strategy to generate and screen *C. elegans* with engineered genomes. Representative procedure for ZFN-, TALEN-, or CRISPR/Cas9-mediated genome editing. First, adult hermaphrodite worm is injected with ZFN, TALEN, or CRISPR/Cas9. Second, an injected worm is transferred onto a new NGM plate in a defined time-course. Finally, worms harboring engineering genomes are identified. In the case of identification of targeted mutations through PCR-based screening such as HMA and Cel-1 assay, individual F1 progeny were picked onto single NGM plate (**a**). Alternatively, observation of a selection marker such as GFP or a known phenotype such as Unc allows us to identify the F1 worms on each plate where an injected worms were self-fertilized (**b**)

At first, the endogenous gene *ben-1* was targeted, because *ben-1*/+mutant has an easily scorable phenotype: mobility on the paralysis-inducing drug benomyl. To overcome prior obstacle for attaining heritable mutations in *C. elegans*, TALEN-encoding mRNAs carrying 5′ and 3′ untranslated regions favorable for germline translation were designed and injected into the *C. elegans* germ-line (Fig. 4.1). For the time course analysis, injected hermaphrodites were moved onto new benomyl plates every 4 hours and the progeny produced within each time window were analyzed independently (Fig. 4.2). TALENs induced many *ben-1* mutations, and 3.5 % of the progeny during a 4-h peak window were heterozygous mutants. TALENs were also applied to other nematode species such as *Caenorhabditis species 9* and *P. pacificus*, which resulted in generating NHEJ-mediated knockout mutant (Lo et al. 2013). As shown in these papers, identification of targeted mutations in *C. elegans* usually depended on observations of fluorescent markers such as green fluorescent protein or visible phenotypes. To apply a TALEN-based strategy to *C. elegans* in more practice, a heteroduplex mobility assay was applied to identify the mutation with no obvious phenotype (Sugi et al. 2014).

The CRISPR/Cas9 system emerged as a conceptually simpler approach that is alternative to TALEN. The first application of CRISPR/Cas9 system for *C. elegans* was based on a simple protocol, in which Friedland et al. constructed two plasmids: one drives expression of transcripts encoding Cas9 fused with SV40 nuclear localization signal (NLS) under the control of *eft-3* promoter that drives transgene expression in the germ-line, and the other is sgRNA complementary to coding sequences in the *unc-119* and *dpy-13* genes from a *U6* small nuclear RNA promoter (Friedland et al. 2013). They microinjected these plasmids into germ-line of *C. elegans* and induced mutations in the two genes with clear mutant phenotypes from the progeny of individual F1 animals. Furthermore, they also targeted the two genes (*klp-12 and Y61A9LA.1*) with no known loss-of-function phenotypes, and were able to generate disruptions of *klp-12 and Y61A9LA.1* in 53/66 (80.3 %) and 13/72 (18.1 %) of the F1 worms screened, respectively. This result revealed that the mutation frequency is high enough that an obvious phenotype or visible selection markers are not necessarily required. Especially, at the *klp-12* locus, 27 out of 80 F1 worms carrying a disruption were homozygous for a single disruption, and the remaining worms were heterozygous for a single disruption or carried two unique mutational alleles.

In late 2013, different research groups developed several useful protocols and demonstrated that heritable changes in nematode genome were induced using CRISPR/Cas9 (Chen et al. 2013b; Chiu et al. 2013; Cho et al. 2013; Frøkjaer-Jensen 2013; Katic and Großhans 2013; Tzur et al. 2013; Waaijers et al. 2013). The protocols were mainly different in the delivery of CRISPR-Cas9 effector complex (DNA, RNA, or protein) and in the identification of mutation (PCR screen, visible markers, or obvious phenotypes). Most groups expressed *C. elegans* codon optimized Cas9 under the control of *eft-3* promoter, whereas Boxem's group used heat-shock promoter to express Cas9 during a limited time window and prevent the deleterious side-effects they observed in the case of use of *eft-3* promoter for Cas9 expression (Waaijers et al. 2013). Other groups used Cas9 and sgRNAs transcribed in vitro (Chiu et al. 2013; Katic and Großhans 2013) or with CRISPR RNAs (crRNAs) and trans-acting crRNAs (tracrRNAs) transcribed separately in vitro (Lo et al. 2013). As a unique protocol, sgRNA was delivered to *C. elegans* through bacterial feeding to achieve gene disruptions in a time- and labor saving manner like the RNAi feeding strategy, which is a powerful technique to deliver gene-specific double-stranded RNAs to major tissues in *C. elegans* (Liu et al. 2014). In this protocol, Cas9 was driven by the *pie-1* promoter, which also drives transgene expression in the germ-line. The transgenic worms carrying *pie-1p::Cas9* DNA as an extrachromosomal array were fed the HT115 bacteria transformed with sgRNA. The mutation rates for *bli* and *dpy* loci in the feeding protocol were 1.40 % and 0.92 %, respectively, whereas the mutation rate of *dpy* in the injection of sgRNA at the low and high concentrations were 7.54 % and 2.29 %, respectively. Despite the low mutation rate, the CRISPR/Cas9 feeding protocol has particular advantages: the continuous feeding of sgRNA bacteria facilitates experimental management, and the protocol can be further modified for large-scale studies, with a worm sgRNA library.

4.2.3 Genome Editing by TALEN- and CRISPR-Mediated Homologous Recombination

A DNA double-strand break induced by CRISPR/Cas9 system can be efficiently repaired by homologous recombination. Therefore, supplying engineered homologous repair templates enables site-directed gene insertion. Dickinson et al. targeted the *nmy-2* gene, which encodes nonmuscle myosin II, and designed a homologous repair template comprising the C-terminal 1.5 kilobases (kb) of *nmy-2* fused in-frame to *gfp*, which was followed by the *nmy-2* 3′ UTR, and *unc-119*(+) selection marker and 1.5 kb of downstream genomic sequence (Dickinson et al. 2013). The *unc-119*(+) gene was flanked by loxP sites so that this gene can be excised by expression of *Cre* recombinase (Fig. 4.3). Expression and localization of the GFP-tagged NMY-2 protein were observed in the cell cortex of embryo, and the insertion of *gfp* at the endogenous locus did not affect *nmy-2* gene function. Several other groups also succeeded

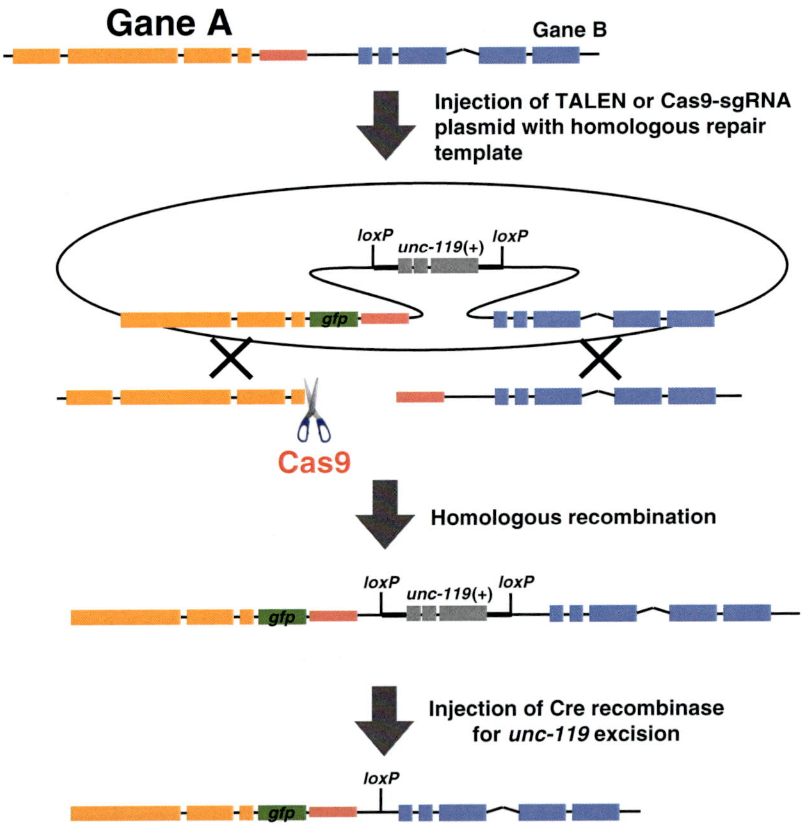

Fig. 4.3 Cas9-mediated homologous recombination in *C. elegans*. Schematic for homologous recombination-mediated *gfp* insertion using TALEN or Cas9. The *unc-119*(+) selection marker in the resulting *gfp*-integrated mutant can be excited by using Cre recombinase. Figure was modified from Dickson et al. (2013)

in TALEN-mediated insertion of the gene encoding the HA tag in *P. pacificus* and in CRISPR/Cas9-based site-directed insertion of the genes encoding GFP and HygR, which enables hygromycin B resistant selection of desired mutants, in *C. elegans* (Chen et al. 2013b; Lo et al. 2013; Tzur et al. 2013). Katic et al. designed an sgRNA targeting a 20-bp sequence in the *daf-2* gene overlapping the mutation *m579*, which causes a missense change in the ligand-binding domain of the insulin receptor homolog DAF-2 (Scott et al. 2002), and demonstrated transgene-instructed gene conversion of the *daf-2(m579)* allele through Cas9-induced double-strand break repair by a wild-type sequence (Katic and Großhans 2013).

4.2.4 Applications of Genome Editing Technology

A most remarkable feature of *C. elegans* genetics is that genes such as a cDNA and fluorescence markers can be expressed in a target cell including individual neurons using a cell-specific promoter identified based on the observation of expression pattern of the promoter-driven GFP (Mello et al. 1991) (Fig. 4.4). This advantage promises to achieve the conditional editing of a targeted genome sequence in *C. elegans*. The conditional editing provides the opportunity to investigate the function of a gene that causes embryonic lethal during larval development in *C. elegans*. In proof-of-principle, Cheng et al. demonstrated that TALEN-mediated nonhereditary knockout could be performed in a cell-specific manner (Cheng et al. 2013). They targeted the

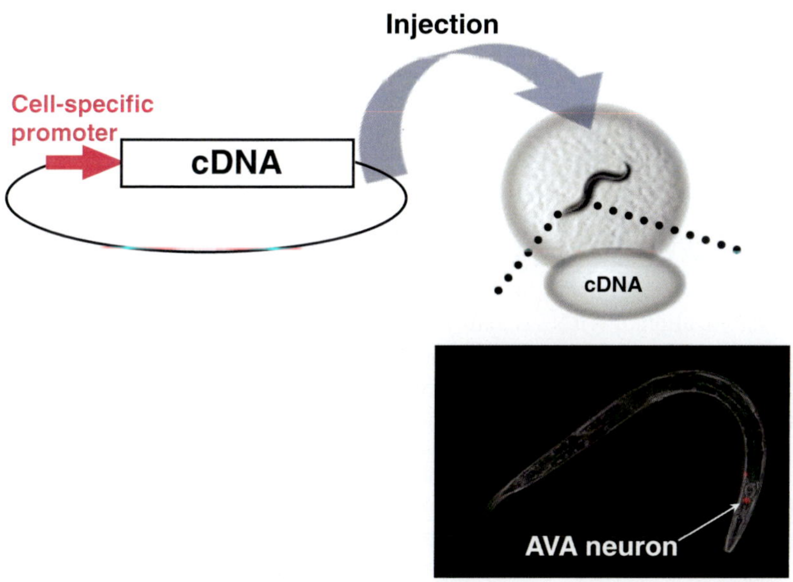

Fig. 4.4 Cell-specific gene expression in *C. elegans*. An example of cell-specific gene expression. A gene of interest is fused with a promoter for expressing transgene in a target neuron. DsRed expressed under the control of the *flp-18* promoter was observed in AVA neurons

cor-1 gene, a worm homolog of the severe combined immunodeficiency (SCID) gene coronin, and the embryonic lethality was caused by a *cor-1* null mutation. Although it has been known that coronins encode a conserved family of actin-binding proteins in mouse (Föger et al. 2006), its function remained unknown in *C. elegans*. Considering that the migration of Q-cell whose asymmetric divisions generate AQR and PQR neurons was actin-dependent processes during *C. elegans* larval development, it was fascinating to investigate the role of *cor-1* gene in Q-cell development. The somatic TALEN for the *col-1* gene was expressed under the control of Q-cell specific *egl-17* promoter. In this transgenic animal, Q-cell descendants, AQR and PQR neurons, moved in the correct directions in *C. elegans* larva; however, their migration distances were substantially reduced compared with those in wild-type animals. Sequence analysis revealed that deletion/insertion mutations by *cor-1* TALENs caused the frameshift in the *cor-1* open reading frame, resulting in the migration defect of Q-cell descendants. Thus, using germ-line transformation with plasmids encoding TALENs under the control of an inducible or cell-specific promoter, effective gene modifications and resulting phenotypes in specific developmental stages and tissues were observed (Cheng et al. 2013). Similar strategy was also developed in CRISPR/Cas9 system using other cell-specific promoters (Liu et al. 2014).

More recently, 'co-CRISPR' strategy was established that facilitates the identification of functional sgRNAs, and enriches for transgenic worms carrying a homologous recombination event (Kim et al. 2014). The authors have found that about half of sgRNAs tested were not effective in practice and expected that an sgRNA targeting the muscle structural gene *unc-22*, which was proven to work effectively and resulted in an easily observable paralyzed twitching phenotype (Moerman and Baillie 1979), would enable efficient identification of animals in which CRISPR/Cas9 system was active. To test this co-CRISPR strategy, they co-injected the *unc-22* sgRNA with a targeted sgRNA. Interestingly, selecting for the twitching phenotype dramatically enriched for animals in which a targeted sgRNA was active.

In accordance with the development of genome editing technique, this is also practically applied to gain biological significance into *C. elegans* (Miki et al. 2014). The targeted inactivation of *paxt-1* gene, which is a subunit of an XRN2 complex that encodes a eukaryotic exoribonuclease to process and degrade various substrates, was achieved by TALEN-mediated genome editing. This inactivation reduced XRN2 protein levels, decreased miRNA turnover activity, and resulted in worm death, suggesting that stabilization of XRN2 is a major function of PAXT-1 in *C. elegans*. Thus, the practical application of genome editing technique opens new avenue to unveil novel roles of genes, which have been not investigated with mutational work so far.

Conclusion

Genome editing techniques have become indispensable tools for interrogating, perturbing, and engineering biological systems in nematode species, like other model organisms. The previously available methods to generate mutations, such as chemical mutagenesis, require laborious procedures that take at least 1 month to conduct large-scale mutagenesis screening. On the other hand, recently developed genome editing techniques allow us to conditionally engineer the targeted mutations using cell-specific promoter. This approach provides an opportunity to cell-specifically investigate the role of a gene of interest, preventing its null mutation from causing embryonic lethal in *C. elegans*.

In other model organisms, TALE and Cas9 can be also converted into a synthetic transcriptional regulator by fusing them to VP16/VP64 for transcriptional activator (Crocker and Stern 2013; Konermann et al. 2013) or to an epigenetic factor such as LSD1 histone demethylase (Mendenhall et al. 2013). In addition, fluorescently tagged Cas9 labeling of specific DNA loci was developed as a powerful live-cell imaging alternative to fluorescence in situ hybridization (FISH) methods (Chen et al. 2013a). These techniques should be highly compatible with *C. elegans* genetics, because it is possible to interrogate the causal roles of a gene expression change and an associated chromatin state at the single cell level in diverse biological processes including development, learning and memory, and disease. In future perspectives, hopefully, the useful genome editing techniques will be combined with the nematode cell-specific genetics toward understating the biological mechanisms at the cellular resolution.

References

Brenner S (1974) The genetics of *Caenorhabditis elegans*. Genetics 77:71–94

C. elegans Deletion Mutant Consortium (2012) Large-scale screening for targeted knockouts in the *Caenorhabditis elegans* genome. G3 (Bethesda) 2:1415–1425

Chen B, Gilbert LA, Cimini BA, Schnitzbauer J, Zhang W, Li G-W, Park J, Blackburn EH, Weissman JS, Qi LS, Huang B (2013a) Dynamic imaging of genomic loci in living human cells by an optimized CRISPR/Cas system. Cell 155:1479–1491

Chen C, Fenk LA, de Bono M (2013b) Efficient genome editing in *Caenorhabditis elegans* by CRISPR-targeted homologous recombination. Nucleic Acids Res 41:e193

Cheng Z, Yi P, Wang X, Chai Y, Feng G, Yang Y, Liang X, Zhu Z, Li W, Ou G (2013) Conditional targeted genome editing using somatically expressed TALENs in *C. elegans*. Nat Biotechnol 31:934–937

Chiu H, Schwartz HT, Antoshechkin I, Sternberg PW (2013) Transgene-free genome editing in *Caenorhabditis elegans* using CRISPR-Cas. Genetics 195:1167–1171

Cho SW, Lee J, Carroll D, Kim J-S, Lee J (2013) Heritable gene knockout in *Caenorhabditis elegans* by direct injection of Cas9-sgRNA ribonucleoproteins. Genetics 195:1177–1180

Crocker J, Stern DL (2013) TALE-mediated modulation of transcriptional enhancers in vivo. Nat Meth 10:762–767

Dickinson DJ, Ward JD, Reiner DJ, Goldstein B (2013) Engineering the *Caenorhabditis elegans* genome using Cas9-triggered homologous recombination. Nat Meth 10:1028–1034

Föger N, Rangell L, Danilenko DM, Chan AC (2006) Requirement for coronin 1 in T lymphocyte trafficking and cellular homeostasis. Science 313:839–842

Friedland AE, Tzur YB, Esvelt KM, Colaiácovo MP, Church GM, Calarco JA (2013) Heritable genome editing in *C. elegans* via a CRISPR-Cas9 system. Nat Meth 10:741–743

Frøkjaer-Jensen C (2013) Exciting prospects for precise engineering of *Caenorhabditis elegans* genomes with CRISPR/Cas9. Genetics 195:635–642

Frøkjaer-Jensen C, Davis MW, Hopkins CE, Newman BJ, Thummel JM, Olesen S-P, Grunnet M, Jorgensen EM (2008) Single-copy insertion of transgenes in *Caenorhabditis elegans*. Nat Genet 40(11):1375–1383

Kage-Nakadai E, Kobuna H, Funatsu O, Otori M, Gengyo-Ando K, Yoshina S, Hori S, Mitani S (2012) Single/low-copy integration of transgenes in *Caenorhabditis elegans* using an ultraviolet trimethylpsoralen method. BMC Biotechnol 12:1

Katic I, Großhans H (2013) Targeted heritable mutation and gene conversion by Cas9-CRISPR in *Caenorhabditis elegans*. Genetics 195:1173–1176

Kim H, Ishidate T, Ghanta KS, Seth M, Conte D, Shirayama M, Mello CC (2014) A Co-CRISPR strategy for efficient genome editing in *Caenorhabditis elegans*. Genetics 197:1069–1080

Konermann S, Brigham MD, Trevino AE, Hsu PD, Heidenreich M, Cong L, Platt RJ, Scott DA, Church GM, Zhang F (2013) Optical control of mammalian endogenous transcription and epigenetic states. Nature 500:472–476

Liu P, Long L, Xiong K, Yu B, Chang N, Xiong J-W, Zhu Z, Liu D (2014) Heritable/conditional genome editing in *C. elegans* using a CRISPR-Cas9 feeding system. Cell Res 24:886–889

Lo T-W, Pickle CS, Lin S, Ralston EJ, Gurling M, Schartner CM, Bian Q, Doudna JA, Meyer BJ (2013) Precise and heritable genome editing in evolutionarily diverse nematodes using TALENs and CRISPR/Cas9 to engineer insertions and deletions. Genetics 195:331–348

Mello CC, Kramer JM, Stinchcomb D, Ambros V (1991) Efficient gene transfer in *C. elegans*: extrachromosomal maintenance and integration of transforming sequences. EMBO J 10: 3959–3970

Mendenhall EM, Williamson KE, Reyon D, Zou JY, Ram O, Joung JK, Bernstein BE (2013) Locus-specific editing of histone modifications at endogenous enhancers. Nat Biotechnol 31:1133–1136

Miki TS, Richter H, Rüegger S, Großhans H (2014) PAXT-1 promotes XRN2 activity by stabilizing it through a conserved domain. Mol Cell 53:351–360

Moerman DG, Baillie DL (1979) Genetic organization in *CAENORHABDITIS ELEGANS*: fine-structure analysis of the unc-22 gene. Genetics 91:95–103

Morton J, Davis MW, Jorgensen EM, Carroll D (2006) Induction and repair of zinc-finger nuclease-targeted double-strand breaks in *Caenorhabditis elegans* somatic cells. Proc Natl Acad Sci U S A 103:16370–16375

Mussolino C, Cathomen T (2012) TALE nucleases: tailored genome engineering made easy. Curr Opin Biotechnol 23:644–650

Robert V, Bessereau J-L (2007) Targeted engineering of the *Caenorhabditis elegans* genome following Mos1-triggered chromosomal breaks. EMBO J 26:170–183

Scott BA, Avidan MS, Crowder CM (2002) Regulation of hypoxic death in *C. elegans* by the insulin/IGF receptor homolog DAF-2. Science 296:2388–2391

Shaye DD, Greenwald I (2011) OrthoList: a compendium of *C. elegans* genes with human orthologs. PLoS One 6:e20085

Sugi T, Sakuma T, Ohtani Y, Yamamoto T (2014) Versatile strategy for isolating transcription activator-like effector nuclease-mediated knockout mutants in *Caenorhabditis elegans*. Dev Growth Differ 56:78–85

Sulston JE, Schierenberg E, White JG, Thomson JN (1983) The embryonic cell lineage of the nematode *Caenorhabditis elegans*. Dev Biol 100:64–119

Tzur YB, Friedland AE, Nadarajan S, Church GM, Calarco JA, Colaiácovo MP (2013) Heritable custom genomic modifications in *Caenorhabditis elegans* via a CRISPR-Cas9 system. Genetics 195:1181–1185

Waaijers S, Portegijs V, Kerver J, Lemmens BBLG, Tijsterman M, Van Den Heuvel S, Boxem M (2013) CRISPR/Cas9-targeted mutagenesis in *Caenorhabditis elegans*. Genetics 195:1187–1191

White JG, Southgate E, Thomson JN, Brenner S (1986) The structure of the nervous system of the nematode *Caenorhabditis elegans*. Philos Trans R Soc Lond B Biol Sci 314:1–340

Wood AJ, Lo T-W, Zeitler B, Pickle CS, Ralston EJ, Lee AH, Amora R, Miller JC, Leung E, Meng X, Zhang L, Rebar EJ, Gregory PD, Urnov FD, Meyer BJ (2011) Targeted genome editing across species using ZFNs and TALENs. Science 333:307

Chapter 5
Highly Efficient Targeted Gene Disruption in the Silkworm, *Bombyx mori*, Using Genome Editing Tools

Takaaki Daimon

Abstract The silkworm, *Bombyx mori*, is a classic model organism in studies of insect genetics and physiology. As *B. mori* and most other lepidopteran species (moths and butterflies) are generally refractory to RNA interference (RNAi), it has been very difficult to conduct loss-of-function studies of genes in lepidopterans. However, recent advances in genome engineering tools, such as zinc finger nucleases (ZFNs), transcription activator-like effector nucleases (TALENs), and the clustered regularly interspaced palindromic repeats/CRISPR-associated (CRISPR/Cas) system, have dramatically changed this situation. Although efficiency of targeted mutagenesis in *B. mori* was very low in an early experiment using ZFNs, recent studies using TALENs and CRISPR/Cas9 have induced highly efficient mutagenesis of the target genes in *B. mori*, even when the mutant phenotype was unknown. Genome editing tools facilitate sophisticated genetic manipulation and breeding of a wide variety of both beneficial and pest insects. This chapter summarizes recent advances in genome editing techniques in *B. mori* and proposes guidelines for experimental design and strategy for successful gene knockout experiments using this species.

Keywords Cas9 • CRISPR • Knockout • Lepidoptera • TALEN

5.1 Introduction

The domesticated silkworm, *Bombyx mori*, is a classic model insect organism in studies of insect physiology and genetics (Tazima 1964, 1978; Goldsmith et al. 2005). *B. mori* is the first lepidopteran insects (the group that includes moths and butterflies) whose whole genome sequence was determined (Mita et al. 2004; Xia et al. 2004; The International Silkworm Genome Consortium 2008). In addition, an

T. Daimon (✉)
National Institute of Agrobiological Sciences,
Owashi 1-2, Tsukuba city, Ibaraki 305-8634, Japan
e-mail: daimontakaaki@affrc.go.jp

T. Yamamoto (ed.), *Targeted Genome Editing Using Site-Specific Nucleases*,
DOI 10.1007/978-4-431-55227-7_5

efficient germline transformation system has been established (Tamura et al. 2000; Uchino et al. 2007). To date, various genetic tools have been developed, such as the GAL4/UAS system (Imamura et al. 2003; Kobayashi et al. 2011), the enhancer trap system (Uchino et al. 2008), a Cre-mediated recombination system (Duan et al. 2013), and an in vivo lipofection/electroporation system (Ando and Fujiwara 2012; Kamimura et al. 2012). These genetic tools, together with genetic resources, including more than 1,000 mutant strains and geographic and commercial races (Tazima 1964, 1978; Banno et al. 2005), facilitate functional studies of genes of interest as well as the molecular breeding of silkworms.

However, there is one disadvantage in B. mori: this species (like most lepidopterans) apparently lacks efficient systemic RNA interference (RNAi) machinery (Terenius et al. 2011; Kobayashi et al. 2012), making it difficult to achieve loss-of-function studies using RNAi in B. mori. Although several research groups successfully conducted RNAi experiments, they adopted special techniques to express double-stranded RNA, such as the use of viral vectors (Uhlirova et al. 2003) and the introduction of transgenes to express RNA hairpins (Subbaiah et al. 2013). Recently, it was shown that injection of short-interfering RNA (siRNA) induces strong gene-silencing effects in B. mori embryos (Yamaguchi et al. 2011). However, such strong effects were observed only when siRNA was injected into embryos. Therefore, *B. mori* genetics has long awaited the establishment of a practical gene targeting system.

Recent rapid advances in genome editing tools, such as zinc finger nucleases (ZFNs) (Urnov et al. 2010), transcription activator-like (TAL) effector nucleases (TALENs) (Joung and Sander 2013), and the clustered regularly interspaced palindromic repeats/CRISPR-associated (CRISPR/Cas) system (Cho et al. 2013; Hwang et al. 2013a, b; Jiang et al. 2013), have enabled efficient gene knockout experiments in a wide variety of model and non-model organisms. Takasu et al. (2010) were the first group to successfully target genes in *B. mori* using genome-editing tools. They used ZFNs to disrupt a phenotypic marker gene, *BmBLOS2*, and they successfully obtained novel mutant alleles, most of which contained short insertions or deletions (indels) (Takasu et al. 2010). However, germline mutation rates in *B. mori* using ZFNs were much lower than those reported in the fruit fly, *Drosophila melanogaster*. Therefore, ZFN-mediated knockout experiments in *B. mori* still require great effort, especially when the goal is to target genes with unknown phenotypes.

Recently, several reports have shown that TALENs and CRISPR/Cas9 can disrupt target genes much more efficiently than ZFNs in *B. mori* (Ma et al. 2012, 2014a, b; Sajwan et al. 2013; Takasu et al. 2013; Wang et al. 2013; Daimon et al. 2014; Liu et al. 2014). Notably, when TALENs are used, germline mutation rates sometimes reach more than 50 % (Takasu et al. 2013; Daimon et al. 2014). Thus, TALENs and CRISPR/Cas9 can now be used as standard tools for gene targeting studies, even when mutant phenotypes are unknown.

In this chapter, I review the recent progress in gene targeting techniques in *B. mori* and suggest guidelines for successful knockout experiments. Targeted gene disruption using engineered nucleases has now become a standard technique that is easy to apply in *B. mori*.

5.2 Targeted Gene Disruption Using TALENs

TALENs consist of a DNA binding domain, which originates from bacterial TAL effectors, and a *Fok*I nuclease domain, which induces double-strand breaks at the target site (Christian et al. 2010; Miller et al. 2011). The highly modular structure of the DNA binding domain of TAL effectors and its simple DNA binding code (Boch et al. 2009; Moscou and Bogdanove 2009) enables researchers to design and assemble custom TALENs with ease. TALENs have been used for targeted gene disruption in many types of organisms, including animals, plants, and yeast (Joung and Sander 2013; Xiao et al. 2013b), and successful gene knockout experiments have been reported in *D. melanogaster* (Liu et al. 2012b), *B. mori* (Ma et al. 2012; Sajwan et al. 2013; Takasu et al. 2013; Daimon et al. 2014), and crickets (Watanabe et al. 2012).

Two research groups reported the first successful example of targeted gene disruption using TALENs (Ma et al. 2012; Sajwan et al. 2013). The efficiency of TALENs greatly varied on the basis of TALEN architectures. TAL effector domain of TALENs used by Sajwan *et al.* (2013) had 287 and 232 amino acid residues at the N- and C-terminal sides of the DNA binding domain, respectively, whereas those used by Ma et al. (2012) had truncations at both sides. As has been reported for other animals (Miller et al. 2011), truncated forms exhibited higher levels of activity in *B. mori*, with germline mutation rates ranging from 0.4 % to 61 %. To further increase the efficiency of TALENs in *B. mori*, Takasu et al. (2013) developed a new TALEN backbone vector optimized for use in *B. mori*. The new vector, pBlue-TAL, is based on the truncated TALEN architecture of Miller et al. (2011) and is codon-optimized for *B. mori*. As pBlue-TAL vector is compatible with the Golden Gate Assembly Kit by Cermak et al. (2011) and other add-on kits, users can easily construct custom TALEN vectors using these kits. TALENs constructed in pBlue-TAL showed surprisingly high efficiency, with mutation rate in G_0 gametes above 70 % (Takasu et al. 2013).

In my experiments, I have targeted more than 12 genes or loci in *B. mori* using TALENs. Figure 5.1 shows the efficiency of single-gene knockout experiments. The work has shown that there is a clear correlation between the ratio of G_0 moths yielding at least one mutant G_1 offspring (% G_0 yielders, Fig. 5.1 *x*-axis) and the germline mutation rates in G_0 yielders (Fig. 5.1 *y*-axis). The values of % G_0 yielders can reach 100 %, and the germline mutation rates can exceed 50 %. Some TALENs showed much lower efficiencies, but even in such cases the germline mutation rates were higher than 5 %, which is still a practical level. Notably, all the TALENs tested so far have successfully mutagenized target genes.

5.3 Targeted Gene Disruption Using the CRISPR/Cas9 System

In the CRISPR/Cas9 system, custom guide RNA is created in cultured cells or embryos to direct site-specific DNA cleavage by the Cas9 endonuclease (Cho et al. 2013; Hwang et al. 2013a, b; Jiang et al. 2013). Several procedures have been

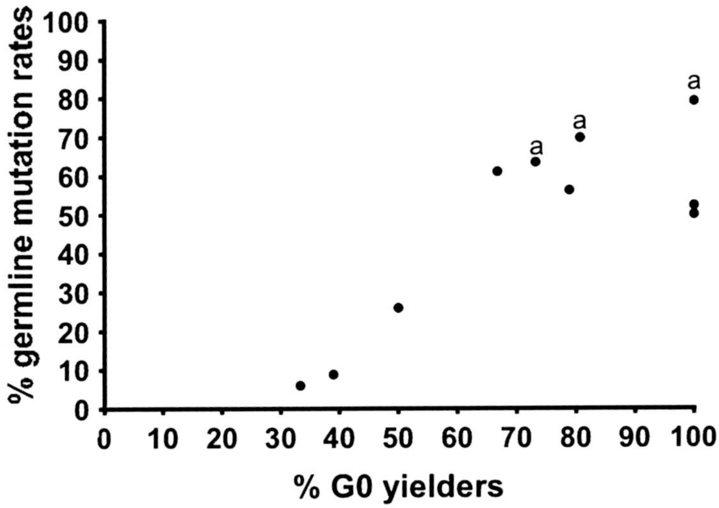

Fig. 5.1 Efficiencies of transcription activator-like effector nucleases (TALEN)-mediated targeted mutagenesis in *Bombyx mori*. The *x*-axis indicates the frequency of G_0 moths that yield at least one G_1 offspring with mutant alleles (% G_0 yielders). The *y*-axis indicates mutation rates in the gametes of G_0 yielders. Each point represents the result of different target genes ($n = 10$). The points indicated by the letter "a" are calculated per G_0 pair, while the others are per individual G_0 moth

developed for using the CRISPR/Cas9 system in vivo. A two-component system has proven to be effective in *D. melanogaster*, in which CRISPR RNA and trans-activating CRISPR RNA are fused into a single RNA called synthetic guide RNA (gRNA), and the Cas9 is expressed from an expression vector (Gratz et al. 2013) or provided as mRNA synthesized in vitro (Bassett et al. 2013; Yu et al. 2013).

The CRISPR/Cas9 system can be also used for targeted mutagenesis in *B. mori*. Wang et al. (2013) were the first to disrupt an endogenous gene of *B. mori* in vivo. They tested two gRNAs, each of which was designed to disrupt the *BmBLOS2* gene. Notably, most (>94 %) of the G_0 larvae that hatched from injected eggs showed the mosaic oily skin phenotype, which was caused by short indels introduced to the target sites. They also performed co-injection of the two gRNAs and found that most (95.5 %) of the G_0 larvae showed the mosaic phenotype, and 35.6 % of G_0 animals yielded at least one germline mutant in the next generation. The target sites of the two gRNAs were ~3.5-kb apart from each other, and ~13.5 % of mutant G_1 larvae screened (7 out of 52) had large deletions of this ~3.5-kb region. These results suggest that CRISPR/Cas9 is a powerful tool for genome manipulation in *B. mori*.

In a similar report, Daimon et al. (2014) also performed gene targeting of the *BmBLOS2* gene using the CRISPR/Cas9 system. However, they achieved much lower efficiency of gene targeting than did Wang et al. (2013), and the germline mutation rate was as low as 0.46 %. The reason for the differences in the efficiencies of the two reports is unclear, but it is likely that the efficiency of CRISPR/Cas9 system is greatly affected by experimental conditions, such as vectors used for the expression of gRNAs and Cas9, doses of RNA injected, and nucleotide sequences of target sites. Further efforts are required to determine the optimal experimental conditions of the CRISPR/Cas9 system for practical use in *B. mori*.

5.4 A Guideline for Gene Targeting Using Genome Editing Tools

5.4.1 Choice of Genome Editing Tools

Although ZFNs, TALENs, and CRISPR/Cas9 are all effective in *B. mori*, TALENs and CRISPR/Cas9 are more promising choices, given their high levels of activity. Before choosing TALENs or CRISPR/Cas9 for genome editing experiments, the following points should be considered. First, construction of TALEN vectors is more complex and requires more time than the construction of gRNA vectors for the CRISPR/Cas9 system (5 vs. 2 days, following standard procedures) (Cermak et al. 2011; Sakuma et al. 2013). Second, TALENs appear to be more robust than CRISPR/Cas9 in *B. mori*. In my experience and that of my colleagues in targeting more than 20 genes and loci, we have rarely failed to generate knockout *B. mori* using TALENs. On the other hand, the efficiency of CRISPR/Cas9 greatly varied depending on the target sites, and some gRNAs did not have high enough activity levels to generate knockouts. Therefore, the most promising method to disrupt a target gene in *B. mori* is TALENs. Third, the CRISPR/Cas9 system can be easily multiplexed by introducing multiple gRNAs. This advantage can be utilized for generating knockouts, as co-injection of several gRNAs for the same target gene will greatly increase the likelihood that gene targeting will be successful. However, care should be taken when multiplex gene targeting is performed because the risk of off-target cleavage greatly increases when multiple gRNAs are introduced simultaneously.

5.4.2 Target Design

With TALENs, the number of repeat-variable di-residue (RVD) modules and the length of the spacer flanked by two TALEN monomers should be determined on the basis of individual TALEN architecture. For example, when TALENs are assembled in the pBlue-TAL vector (Takasu et al. 2013), the parameters of 15–20 RVD repeats and 14- to 16-bp spacers are best. Several public tools are available to help find target sites and design RVD repeats (e.g., TALEN targeter, https://tale-nt.cac.cornell.edu/) (Doyle et al. 2012). Several guidelines have been proposed for the design of efficient TALENs, and I recommend following these guidelines as long as the software identifies a binding site at the desired location. Designed RVD repeats should then be subjected to an off-target search against the whole genome sequence of *B. mori* to minimize the possibility of performing off-target cleavage of the genome. Several public tools have been developed for this purpose, such as the Paired Target Finder (Doyle et al. 2012) and TALENoffer (Grau et al. 2013). Target sites of TALENs can be flexibly designed by changing the number of RVD modules and the spacer length. Therefore, most genomic regions can be targeted.

When one uses CRISPR/Cas9, it is important to design gRNA sequences that are specific to target sites. Cas9 can be introduced using a pre-made Cas9 expression vector. The targeting sequences of CRISPR/Cas9 are limited to 23 bp and should follow the

N20-NGG rule. When gRNAs are expressed by the T7 or U6 promoters, the first base of the target site should be G (G-N19-NGG). However, if the first base of the target site is not G, an additional G can simply be added to the 5'-end (G-N20-NGG) because mismatches at the 5'-end of the target site are tolerated in the CRISPR/Cas9 system. To avoid off-target cleavage of the genome, the specificity of the target sequence should be carefully examined with a homology search or by using a tool for mapping short reads to reference genomes, such as Bowtie (Langmead et al. 2009).

5.4.3 Construction of TALEN or gRNA Vectors

For construction of TALEN vectors, Golden Gate Assembly (GGA), described by Cermak et al. (2011), is best on a single-lab scale. Sakuma et al. (2013) developed an add-on kit for GGA that greatly improved the efficiency of construction of TALEN vectors. Both vector kits are compatible with the pBlueTAL vector, and these vectors are available from Addgene. It is very important to follow the exact procedures described in Cermak et al. (2011) and Sakuma et al. (2013), including use of the same kits and enzymes as the ones described there. Construction of TALEN vectors is usually complete in 4–5 days. As almost all the TALENs my colleagues and I have constructed have showed high levels of activity, sufficient to obtain germline mutants, there is no need to perform validation of the assembled vector before injection. Without functional validation, cleavage activity levels can be evaluated after injection by observing the mosaic phenotype of G_0 individuals (if observable) or by conducting molecular diagnostics, such as a CEL-I or T7 endonuclease I (T7EI) assay, using genomic DNAs of injected G_0 animals.

For construction of gRNA vectors, various "empty" gRNA vectors have been developed and deposited in Addgene. Some vectors are designed for in vitro transcription of gRNAs from the T7 promoter and thus enable the preparation of gRNAs in vitro (Hwang et al. 2013b), whereas some vectors have the U6 promoter and thus enable transcription of gRNAs when they are introduced into cells or embryos (Gratz et al. 2013). Most of these plasmids have unique restriction sites that allow easy and rapid assembly of gRNA vectors by ligating pairs of short, annealed oligonucleotides into the vector backbone. There are also various kinds of Cas9 vectors: like gRNA vectors, some enable in vitro synthesis of Cas9 mRNA, and some are used for transfection into cells or injection into embryos. As these Cas9 vectors are in many cases optimized for the recipient organisms (e.g., *D. melanogaster*, humans, zebrafish), they would require further optimization in order to increase the cleavage activity in *B. mori*.

5.4.4 Mating Scheme

As one generation of *B. mori* takes ~50 days, it is important to make a well-organized plan to optimize the likelihood of achieving successful targeting in the first trial. This is particularly true when the goal is to target an autosomal gene with an unknown phenotype, because an induced mutation should be "tracked" solely on the basis of molecular diagnostics. Here, I propose a mating scheme and screening strategy for such cases (Fig. 5.2).

In the case of TALENs, the percentage of G_0 yielders sometimes reaches 100 % if target genes do not cause lethality in mosaic G_0 animals (Ma et al. 2012; Takasu et al. 2013; Daimon et al. 2014), but the percentage significantly decreases if mosaic

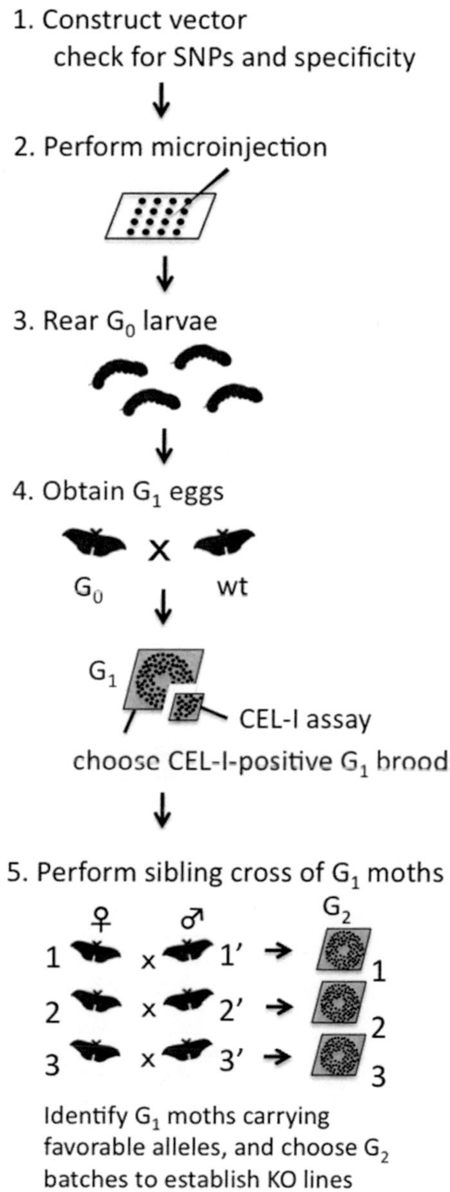

Fig. 5.2 An example of CEL-I assay in G_1. Neonate larvae were collected from each G_1 brood, and 5–10 individuals were pooled for DNA extraction. After PCR was complete, the products were treated with CEL-I. Target sites of transcription activator-like effector nucleases (TALENs) are located at the centers of the amplicons (~300 bp); therefore, ~150-bp fragments were observed when small indels were induced (indicated by an *arrow*). *Asterisks* indicate the presence of large deletion alleles induced by TALENs

animals experience high mortality (Daimon et al. 2014). Therefore, more eggs should be injected when lethality in mosaic G_0 animals is expected.

I strongly recommend crossing G_0 moths with parental or wild-type (wt) strains and avoiding sibling G_0 crosses. As germline mutation rates often exceed 50 % in *B. mori*, sibling G_0 crosses "jumble" the genotype of G_1 animals and make it very difficult to establish a line carrying a single, defined mutant allele in the subsequent generations.

After G_1 eggs have been produced, each G_1 brood should be checked for the presence of mutant alleles. As mutant phenotypes are usually not observable in the G_1 generation, this screening should be done by molecular diagnostics, such as a CEL-I assay or T7E1 assay. I usually sample ~40 neonate larvae or eggs from each brood and investigate the presence of mutant alleles using a CEL-I assay. Samples from the same brood can be mixed in a single tube, but I recommend dividing them into several tubes (5–10 larvae/tube) for DNA extraction and subsequent CEL-I assay so that the frequency of germline mutations can be roughly estimated in each brood and compared among broods. Figure 5.3 shows an example of CEL-I assay in a G_1 population. In this experiment, genomic DNA from 5–10 neonate larvae were pooled in a single polymerase chain reaction (PCR) analysis, and the PCR products were digested with CEL-I. Cleaved fragments shown in Fig. 5.1 indicate that at least one individual in the DNA pool carried a mutant allele. As many types of mutant alleles can be recovered from a single G_0 founder, 4–6 G_1 broods are generally sufficient to recover null alleles. In rare cases, large indels can be found (e.g., brood #25 in Fig. 5.1). As such large indels do not require DNA sequencing for genotyping, G_1 broods that have such mutations should be chosen for the subsequent studies.

In the G_1 generation, 50–100 larvae per G_1 brood are reared and subjected to sibling mating to obtain G_2 eggs. After the crossing has been performed, each G_1 moth is given an ID number and subjected to CEL-I assay to identify CEL-I-positive G_1 moths. Next, the nucleotide sequences of induced mutations should be determined. As CEL-I-positive G_1 moths are considered to be heterozygous for the target gene, PCR products cannot be sequenced directly and should be subcloned into a cloning vector. I usually sequence 8–10 clones per individual, because PCR reactions yield three types of PCR product: wt/wt and mutant/mutant homoduplexes and a wt/mutant heteroduplex. Finally, G_2 broods carrying favorable mutant alleles are chosen to establish mutant lines.

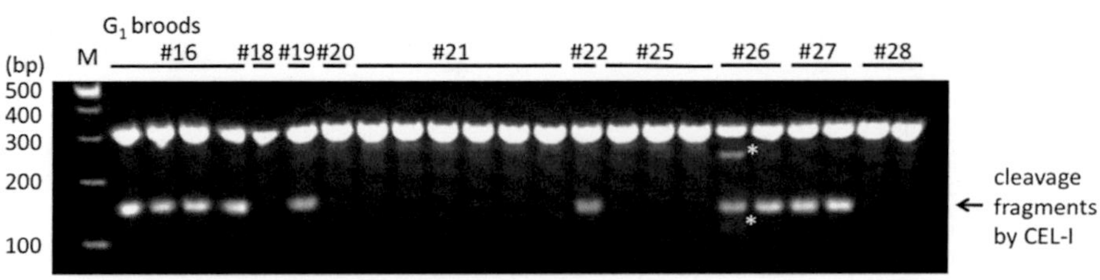

Fig. 5.3 A scheme for gene targeting experiments in *Bombyx mori*. A mating scheme and screening strategy for gene targeting in *B. mori* is proposed. See main text for details

5.4.5 *Common Problems*

Several problems are common in gene targeting experiments, including (i) high lethality in mosaic G_0 animals, (ii) inter- and intra-strain nucleotide polymorphism, and (iii) off-target effects.

5.4.5.1 High Mortality in G_0 Animals

If a target gene causes high lethality in mosaic G_0 animals, it is very difficult to establish germline mutants. Unfortunately, there is no reliable method for circumventing this problem. However, lowering the mutation rates induced by TALENs or CRISPR/Cas9 would be a helpful approach. This can be done by changing the injection conditions (e.g., reducing the doses of RNA injected), using new TALENs that do not follow some of the design guidelines (i.e., spacer length and/or base composition of binding sites), or changing Cas9 expression vectors.

5.4.5.2 Inter- and Intra-Strain Nucleotide Polymorphisms

Inter- and intra-strain nucleotide polymorphisms are problematic in screening of induced mutations because they generate false-positive fragments in CEL-I and T7E1 assays. The whole genome sequence of *B. mori* was determined in an inbred strain, p50T/Dazao (Mita et al. 2004; Xia et al. 2004; The International Silkworm Genome Consortium 2008); therefore, when injection is performed in other strains, it is important to check for the presence of nucleotide polymorphisms before designing TALENs or gRNAs. Special care should also be taken regarding intra-strain polymorphisms when non-isogenic strains are used as recipients (e.g., the *pnd w-1* strain, other hybrid or geographic races).

5.4.5.3 Off-Target Effects

Off-target effects of TALENs and CRISPR/Cas9 have been extensively studied in mammalian systems (Hockemeyer et al. 2011; Pattanayak et al. 2011). Fu et al. (2013) have suggested that in human cells, off-target mutations are induced by CRISPR/Cas9 much more frequently than when ZFNs and TALENs are used. Although it is unknown whether the same is true in insects, off-target effects should be carefully considered. After mutant *B. mori* lines have been established, potential off-target sites should be checked for the presence of unintended mutations. These potential off-target sites can be computationally predicted by public tools such as Paired Target Finder (Doyle et al. 2012) and TALENoffer (Grau et al. 2013). The most reliable method to rule out the possibility that observed phenotypes in knockout lines are caused by disruption of off-target sites is to perform transgenic rescue experiments, although this approach is not practical considering the long generation time of *B. mori*.

One plausible alternative method is to outcross established mutant lines with another strain to check the genetic linkage between a mutant phenotype and the genotype of a target locus in the F_2 generation or a backcross population.

5.5 Further Applications of TALENs and CRISPR/Cas9

TALENs and CRISPR/Cas9 can be used not only to target a single gene but also to perform more sophisticated forms of genome editing, such as inducing large deletions in the genome and gene knockin.

5.5.1 Large Deletions

Large deletions and inversions of the genome can be induced by two pairs of TALENs or CRISPR/Cas9. For example, heritable mutations constituting up to a 122-kb deletion or a 43.8-kb inversion have been successfully induced in zebrafish using TALENs (Xiao et al. 2013a). This approach also works well in *B. mori*. Ma et al. (2012) demonstrated that co-injection of two pairs of TALENs induced large (~800 bp) heritable deletions in the genomic region that was flanked by target sites of each TALEN pair. Similarly, Wang et al. (2013) showed that ~3.5-kb heritable deletion was induced by injecting two gRNAs together with Cas9 mRNA. I have also succeeded in inducing ~4.5-kb deletions in the genomic region that was flanked by two TALEN pairs (Daimon et al., unpublished). Recently, Ma et al. (2014a) showed that ~8.9 Mb segment of the Z chromosome can be deleted or inverted in embryos injected with two pairs of TALENs, although it is unclear whether these deletion and inversion alleles are heritable. The efficiency of large deletions/inversions is lower than that of a single gene targeting event (Xiao et al. 2013a), but it appears to be practical as long as the deletion size is up to several kilobases.

When large deletions are induced, unpredictable small indels are usually generated at the breakpoint junctions. Ma et al. (2014b) showed that the junction of the two breakpoints can be precisely "fixed" by co-injecting single-stranded oligonucleotides (ssODNs) as a repair template together with TALEN mRNA. This approach is useful when large deletions and inversions must be induced in a precise manner.

5.5.2 Gene Knockin

When ZFNs, TALENs, or CRISPR/Cas9 are introduced with a targeting donor vector that has left and right flanking homology arms, induced double-strand breaks can be repaired using the homology-directed repair (HDR) system, enabling precise modifications of target loci. This approach is successful in many cells and organisms, including mammalian cultured cells, mice, rats, zebrafish, *D. melanogaster*, and plants

(Sakuma and Woltjen 2014). However, the efficiency of gene knockin mediated by HDR appears to be very low in *B. mori*. In a green fluorescent protein (GFP) knockin experiment using ZFNs in *B. mori*, only 0.0085 % of G_1 larvae (1 out of 11,770 larvae) carried the GFP knockin allele (Takasu et al. unpublished). As somatic GFP mosaics were observed with a relatively high frequency (9.4 %), it is likely that HDR activity is low in germline cells of *B. mori*.

One promising approach to increase the frequency of HDR is to disrupt genes involved in the non-homologous end-joining (NHEJ) pathway. Indeed, the frequency of HDR after double-strand breaks is greatly increased in *D. melanogaster* mutants that lack a functional DNA ligase IV (lig4), an essential component of NHEJ (Beumer et al. 2008). Ma et al. (2014a, 2014b) used CRISPR/Cas9 to establish knockout B. mori strains lacking a functional Ku70 gene, which is required for the NHEJ pathway, and found that HDR efficiencies were increased in embryos of *Ku70* mutants compared with those of wt embryos. Although the efficiency of knockin events in germline cells was not evaluated by Ma et al. (2014a, 2014b), these *Ku70* mutant strains might be useful for knockin experiments in *B. mori*.

For knockin experiments, ssODNs can be also used as targeting donors to direct DNA repair. This approach allows only subtle modifications (e.g., substitutions of several nucleotides), in contrast with conventional methods that use long donor plasmids (e.g., insertion of several kilobases). The efficiency of genome modifications directed by ssODNs is generally higher than that of modifications directed by targeting plasmids (Sakuma and Woltjen 2014). Therefore, it is interesting to examine the efficiency of knockin using ssODN donors in *B. mori*.

5.5.3 Other Applications

Many variants of TALEN and CRISPR/Cas systems have been developed for various other purposes, in addition to inducing double-strand breaks at the target sites. For example, activation/suppression of genes of interest can be done using TALE activators/repressors and Cas9 activators/repressors, in which a TALE domain or a catalytically inactive Cas9 (dCas9) is fused with an activator/repressor peptide (Cong et al. 2012; Crocker and Stern 2013; Konermann et al. 2013; Mali et al. 2013; Qi et al. 2013). In addition, specific genomic loci can be visualized by TALEs or dCas9 proteins fused with fluorescent proteins (Chen et al. 2013; Miyanari et al. 2013). These new variants will facilitate further genetic manipulation of *B. mori*.

5.6 Future Perspectives

Recent progress in genome editing technologies has initiated a new era for *B. mori* molecular genetics (Daimon et al. 2014). Since the germline transformation of *B. mori* was achieved using transposon vectors, transgenic silkworms have been used as “factories” to produce recombinant proteins of economical and pharmaceutical importance, such as human collagen (Tomita et al. 2003). Other applications for silkworm transgenesis include engineering resistance to insect pathogens (Subbaiah et al. 2013). Genome editing techniques will greatly facilitate the sophisticated molecular breeding of *B. mori* for use in sericulture and biotechnology. For example, desirable traits could be directly introduced to economically valuable races using genome-editing tools. Examples of such traits include resistance to pathogens and pesticides, increased cocoon yields, and the capacity to be genetically sexed. Furthermore, the productivity of recombinant proteins in cocoon shells will be greatly increased if the transgene is integrated into the silk protein gene locus, and new silk fabric can be produced if the coding sequence of the transgene is fused in-flame with the silk protein genes.

Genome editing technologies will also facilitate the genetic engineering of insect pests of agricultural and medical importance. As TALENs and/or CRISPR/Cas9 are considered to be active in most insect species, the only technical barrier for genome editing is the establishment of a delivery system (e.g., injection or electroporation) of RNA or DNA to the early embryo. Germline transformation systems using transposons have been developed for many insects (Handler 2002), including crickets (Nakamura et al. 2010), mosquitoes (Kokoza et al. 2001), flies (Handler and McCombs 2000), wasps (Sumitani et al. 2003), beetles (Lorenzen et al. 2003), and butterflies (Marcus et al. 2004) and moths (Peloquin et al. 2000; Tamura et al. 2000; Ferguson et al. 2011; Liu et al. 2012a; Martins et al. 2012). Therefore, these systems would be helpful guides to establishing a gene delivery system in a new recipient species.

In lepidopterans, successful gene targeting has been reported so far only in two species, *B. mori* (Daimon et al. 2014) and the monarch butterfly (Merlin et al. 2013). In species in which germline transformation systems have been established, targeted gene disruption would not be challenging. These species include the squinting bush brown butterfly (*Bicyclus anynana*) (Marcus et al. 2004), the pink bollworm (*Pectinophora gossypiella*) (Peloquin et al. 2000), the Asian corn borer (*Ostrinia furnacalis*) (Liu et al. 2012a), the codling moth (*Cydia pomonella*) (Ferguson et al. 2011), and the diamondback moth (*Plutella xylostella*) (Martins et al. 2012). As lepidopterans are major agricultural pests, genome editing of lepidopteran pests will provide new forms of pest control beyond the classical techniques based on transgenic sterile insects (Beech et al. 2012).

In conclusion, genome editing tools will facilitate the sophisticated genetic manipulation and breeding of a wide variety of both beneficial and pest insects, thereby greatly impact affecting both basic and applied research.

References

Ando T, Fujiwara H (2012) Electroporation-mediated somatic transgenesis for rapid functional analysis in insects. Development 140:454–458

Banno Y, Fujii H, Kawaguchi Y, Yamamoto K, Nishikawa K, Nishisaka A, Tamura K, Eguchi S (2005) A guide to the silkworm mutants 2005-gene name and gene. Institute of Genetic Resorces, Kyushu University, Fukuoka, Japan, Silkworm Genetics Division

Bassett AR, Tibbit C, Ponting CP, Liu JL (2013) Highly efficient targeted mutagenesis of *Drosophila* with the CRISPR/Cas9 system. Cell Rep 4:220–228

Beech CJ, Koukidou M, Morrison NI, Alphey L (2012) Genetically modified insects: science, use, status and regulation. Collection Biosafety Rev 6:66–124

Beumer KJ, Trautman JK, Bozas A, Liu JL, Rutter J, Gall JG, Carroll D (2008) Efficient gene targeting in *Drosophila* by direct embryo injection with zinc-finger nucleases. Proc Natl Acad Sci U S A 105:19821–19826

Boch J, Scholze H, Schornack S, Landgraf A, Hahn S, Kay S, Lahaye T, Nickstadt A, Bonas U (2009) Breaking the code of DNA binding specificity of TAL-type III effectors. Science 326:1509–1512

Cermak T, Doyle EL, Christian M, Wang L, Zhang Y, Schmidt C, Baller JA, Somia NV, Bogdanove AJ, Voytas DF (2011) Efficient design and assembly of custom TALEN and other TAL effector-based constructs for DNA targeting. Nucleic Acids Res 39:e82

Chen B, Gilbert LA, Cimini BA, Schnitzbauer J, Zhang W, Li GW, Park J, Blackburn EH, Weissman JS, Qi LS, Huang B (2013) Dynamic imaging of genomic loci in living human cells by an optimized CRISPR/Cas system. Cell 155:1479–1491

Cho SW, Kim S, Kim JM, Kim JS (2013) Targeted genome engineering in human cells with the Cas9 RNA-guided endonuclease. Nat Biotechnol 31:230–232

Christian M, Cermak T, Doyle EL, Schmidt C, Zhang F, Hummel A, Bogdanove AJ, Voytas DF (2010) Targeting DNA double-strand breaks with TAL effector nucleases. Genetics 186:757–761

Cong L, Zhou R, Kuo YC, Cunniff M, Zhang F (2012) Comprehensive interrogation of natural TALE DNA-binding modules and transcriptional repressor domains. Nat Commun 3:968

Crocker J, Stern DL (2013) TALE-mediated modulation of transcriptional enhancers in vivo. Nat Methods 10:762–767

Daimon T, Kiuchi T, Takasu Y (2014) Recent progress in genome engineering techniques in the silkworm, *Bombyx mori*. Dev Growth Differ 56:14–25

Doyle EL, Booher NJ, Standage DS, Voytas DF, Brendel VP, Vandyk JK, Bogdanove AJ (2012) TAL Effector-Nucleotide Targeter (TALE-NT) 2.0: tools for TAL effector design and target prediction. Nucleic Acids Res 40 (Web Server issue):W117–122

Duan J, Xu H, Ma S, Guo H, Wang F, Zhao P, Xia Q (2013) Cre-mediated targeted gene activation in the middle silk glands of transgenic silkworms (*Bombyx mori*). Transgenic Res 22:607–619

Ferguson HJ, Neven LG, Thibault ST, Mohammed A, Fraser M (2011) Genetic transformation of the codling moth, *Cydia pomonella* L., with piggyBac EGFP. Transgenic Res 20:201–214

Fu Y, Foden JA, Khayter C, Maeder ML, Reyon D, Joung JK, Sander JD (2013) High-frequency off-target mutagenesis induced by CRISPR-Cas nucleases in human cells. Nat Biotechnol 31:822–826

Goldsmith MR, Shimada T, Abe H (2005) The genetics and genomics of the silkworm, *Bombyx mori*. Annu Rev Entomol 50:71–100

Gratz SJ, Cummings AM, Nguyen JN, Hamm DC, Donohue LK, Harrison MM, Wildonger J, O'Connor-Giles KM (2013) Genome engineering of *Drosophila* with the CRISPR RNA-guided Cas9 nuclease. Genetics 194:1029–1035

Grau J, Boch J, Posch S (2013) TALENoffer: genome-wide TALEN off-target prediction. Bioinformatics 29:2931

Handler AM (2002) Use of the *piggyBac* transposon for germ-line transformation of insects. Insect Biochem Mol Biol 32:1211–1220

Handler AM, McCombs SD (2000) The *piggyBac* transposon mediates germ-line transformation in the oriental fruit fly and closely related elements exist in its genome. Insect Mol Biol 9:605–612

Hockemeyer D, Wang H, Kiani S, Lai CS, Gao Q et al (2011) Genetic engineering of human pluripotent cells using TALE nucleases. Nat Biotechnol 29:731–734

Hwang WY, Fu Y, Reyon D, Maeder ML, Kaini P, Sander JD, Joung JK, Peterson RT, Yeh JR (2013a) Heritable and precise zebrafish genome editing using a CRISPR-Cas system. PLoS One 8:e68708

Hwang WY, Fu Y, Reyon D, Maeder ML, Tsai SQ, Sander JD, Peterson RT, Yeh JR, Joung JK (2013b) Efficient genome editing in zebrafish using a CRISPR-Cas system. Nat Biotechnol 31:227–229

Imamura M, Nakai J, Inoue S, Quan GX, Kanda T, Tamura T (2003) Targeted gene expression using the GAL4/UAS system in the silkworm *Bombyx mori*. Genetics 165:1329–1340

Jiang W, Bikard D, Cox D, Zhang F, Marraffini LA (2013) RNA-guided editing of bacterial genomes using CRISPR-Cas systems. Nat Biotechnol 31:233–239

Joung JK, Sander JD (2013) TALENs: a widely applicable technology for targeted genome editing. Nat Rev Mol Cell Biol 14:49–55

Kamimura M, Saito H, Niwa R, Niimi T, Toyoda K, Ueno C, Kanamori Y, Shimura S, Kiuchi M (2012) Fungal ecdysteroid-22-oxidase, a new tool for manipulating ecdysteroid signaling and insect development. J Biol Chem 287:16488–16498

Kobayashi I, Kojima K, Uchino K, Sezutsu H, Iizuka T, Tatematsu KI, Yonemura N, Tanaka H, Yamakawa M, Ogura E, Kamachi Y, Tamura T (2011) An efficient binary system for gene expression in the silkworm, *Bombyx mori*, using Gal4 variants. Arch Insect Biochem Physiol 76:195–210

Kobayashi I, Tsukioka H, Komoto N, Uchino K, Sezutsu H, Tamura T, Kusakabe T, Tomita S (2012) SID-1 protein of *Caenorhabditis elegans* mediates uptake of dsRNA into *Bombyx* cells. Insect Biochem Mol Biol 42:148–154

Kokoza V, Ahmed A, Wimmer EA, Raikhel AS (2001) Efficient transformation of the yellow fever mosquito *Aedes aegypti* using the *piggyBac* transposable element vector pBac[3xP3-EGFP afm]. Insect Biochem Mol Biol 31:1137–1143

Konermann S, Brigham MD, Trevino A, Hsu PD, Heidenreich M, Le C, Platt RJ, Scott DA, Church GM, Zhang F (2013) Optical control of mammalian endogenous transcription and epigenetic states. Nature 500:472–476

Langmead B, Trapnell C, Pop M, Salzberg SL (2009) Ultrafast and memory-efficient alignment of short DNA sequences to the human genome. Genome Biol 10:R25

Liu D, Yan S, Huang Y, Tan A, Stanley DW, Song Q (2012a) Genetic transformation mediated by *piggyBac* in the Asian corn borer, *Ostrinia furnacalis* (Lepidoptera: Crambidae). Arch Insect Biochem Physiol 80:140–150

Liu J, Li C, Yu Z, Huang P, Wu H, Wei C, Zhu N, Shen Y, Chen Y, Zhang B, Deng WM, Jiao R (2012b) Efficient and specific modifications of the *Drosophila* genome by means of an easy TALEN strategy. J Genet Genomics 39:209–215

Liu Y, Ma S, Wang X, Chang J, Gao J, Shi R, Zhang J, Lu W, Zhao P, Xia Q (2014) Highly efficient multiplex targeted mutagenesis and genomic structure variation in *Bombyx mori* cells using CRISPR/Cas9. Insect Biochem Mol Biol 49:35–42

Lorenzen MD, Berghammer AJ, Brown SJ, Denell RE, Klingler M, Beeman RW (2003) *piggyBac*-mediated germline transformation in the beetle *Tribolium castaneum*. Insect Mol Biol 12:433–440

Ma S, Zhang S, Wang F, Liu Y, Xu H, Liu C, Lin Y, Zhao P, Xia Q (2012) Highly efficient and specific genome editing in silkworm using custom TALENs. PLoS One 7:e45035

Ma S, Chang J, Wang X, Liu Y, Zhang J, Lu W, Gao J, Shi R, Zhao P, Xia Q (2014a) CRISPR/Cas9 mediated multiplex genome editing and heritable mutagenesis of BmKu70 in *Bombyx mori*. Sci Rep 4:4489

Ma S, Wang X, Liu Y, Gao J, Zhang S, Shi R, Chang J, Zhao P, Xia Q (2014b) Multiplex genomic structure variation mediated by TALEN and ssODN. BMC Genomics 15:41

Mali P, Aach J, Stranges PB, Esvelt KM, Moosburner M, Kosuri S, Yang L, Church GM (2013) CAS9 transcriptional activators for target specificity screening and paired nickases for cooperative genome engineering. Nat Biotechnol 31:833–838

Marcus JM, Ramos DM, Monteiro A (2004) Germline transformation of the butterfly *Bicyclus anynana*. Proc Biol Sci 271(Suppl 5):S263–S265

Martins S, Naish N, Walker AS, Morrison NI, Scaife S, Fu G, Dafa'alla T, Alphey L (2012) Germline transformation of the diamondback moth, *Plutella xylostella* L., using the *piggyBac* transposable element. Insect Mol Biol 21:414–421

Merlin C, Beaver LE, Taylor OR, Wolfe SA, Reppert SM (2013) Efficient targeted mutagenesis in the monarch butterfly using zinc-finger nucleases. Genome Res 23:159–168

Miller JC, Tan S, Qiao G, Barlow KA, Wang J, Xia DF, Meng X, Paschon DE, Leung E, Hinkley SJ, Dulay GP, Hua KL, Ankoudinova I, Cost GJ, Urnov FD, Zhang HS, Holmes MC, Zhang L, Gregory PD, Rebar EJ (2011) A TALE nuclease architecture for efficient genome editing. Nat Biotechnol 29:143–148

Mita K, Kasahara M, Sasaki S, Nagayasu Y, Yamada T, Kanamori H, Namiki N, Kitagawa M, Yamashita H, Yasukochi Y, Kadono-Okuda K, Yamamoto K, Ajimura M, Ravikumar G, Shimomura M, Nagamura Y, Shin IT, Abe H, Shimada T, Morishita S, Sasaki T (2004) The genome sequence of silkworm, *Bombyx mori*. DNA Res 11:27–35

Miyanari Y, Ziegler-Birling C, Torres-Padilla ME (2013) Live visualization of chromatin dynamics with fluorescent TALEs. Nat Struct Mol Biol 20:1321–1324

Moscou MJ, Bogdanove AJ (2009) A simple cipher governs DNA recognition by TAL effectors. Science 326:1501

Nakamura T, Yoshizaki M, Ogawa S, Okamoto H, Shinmyo Y, Bando T, Ohuchi H, Noji S, Mito T (2010) Imaging of transgenic cricket embryos reveals cell movements consistent with a syncytial patterning mechanism. Curr Biol 20:1641–1647

Pattanayak V, Ramirez CL, Joung JK, Liu DR (2011) Revealing off-target cleavage specificities of zinc-finger nucleases by in vitro selection. Nat Methods 8:765–770

Peloquin JJ, Thibault ST, Staten R, Miller TA (2000) Germ-line transformation of pink bollworm (Lepidoptera: gelechiidae) mediated by the *piggyBac* transposable element. Insect Mol Biol 9:323–333

Qi LS, Larson MH, Gilbert LA, Doudna JA, Weissman JS, Arkin AP, Lim WA (2013) Repurposing CRISPR as an RNA-guided platform for sequence-specific control of gene expression. Cell 152:1173–1183

Sajwan S, Takasu Y, Tamura T, Uchino K, Sezutsu H, Zurovec M (2013) Efficient disruption of endogenous *Bombyx* gene by TAL effector nucleases. Insect Biochem Mol Biol 43:17–23

Sakuma T, Woltjen K (2014) Nuclease-mediated genome editing: at the front-line of functional genomics technology. Dev Growth Differ 56:2–13

Sakuma T, Hosoi S, Woltjen K, Suzuki K, Kashiwagi K, Wada H, Ochiai H, Miyamoto T, Kawai N, Sasakura Y, Matsuura S, Okada Y, Kawahara A, Hayashi S, Yamamoto T (2013) Efficient TALEN construction and evaluation methods for human cell and animal applications. Genes Cells 18:315–326

Subbaiah EV, Royer C, Kanginakudru S, Satyavathi VV, Babu AS, Sivaprasad V, Chavancy G, Darocha M, Jalabert A, Mauchamp B, Basha I, Couble P, Nagaraju J (2013) Engineering silkworms for resistance to baculovirus through multigene RNA interference. Genetics 193:63–75

Sumitani M, Yamamoto DS, Oishi K, Lee JM, Hatakeyama M (2003) Germline transformation of the sawfly, *Athalia rosae* (Hymenoptera: Symphyta), mediated by a *piggyBac*-derived vector. Insect Biochem Mol Biol 33:449–458

Takasu Y, Kobayashi I, Beumer K, Uchino K, Sezutsu H, Sajwan S, Carroll D, Tamura T, Zurovec M (2010) Targeted mutagenesis in the silkworm *Bombyx mori* using zinc finger nuclease mRNA injection. Insect Biochem Mol Biol 40:759–765

Takasu Y, Sajwan S, Daimon T, Osanai-Futahashi M, Uchino K, Sezutsu H, Tamura T, Zurovec M (2013) Efficient TALEN construction for *Bombyx* gene targeting. PLoS One 18:e73458

Tamura T, Thibert C, Royer C, Kanda T, Abraham E, Kamba M, Komoto N, Thomas JL, Mauchamp B, Chavancy G, Shirk P, Fraser M, Prudhomme JC, Couble P (2000) Germline transformation of the silkworm *Bombyx mori* L. Using a *piggyBac* transposon-derived vector. Nat Biotechnol 18:81–84

Tazima Y (1964) The genetics of the silkworm. Logos Press, London

Tazima Y (1978) The silkworm: an important tool. Kodansha, Tokyo

Terenius O, Papanicolaou A, Garbutt JS, Eleftherianos I, Huvenne H, Kanginakudru S, Albrechtsen M, An C, Aymeric JL, Barthel A, Bebas P, Bitra K, Bravo A, Chevalier F, Collinge DP, Crava CM, de Maagd RA, Duvic B, Erlandson M, Faye I, Felfoldi G, Fujiwara H, Futahashi R, Gandhe

AS, Gatehouse HS, Gatehouse LN, Giebultowicz JM, Gomez I, Grimmelikhuijzen CJ, Groot AT, Hauser F, Heckel DG, Hegedus DD, Hrycaj S, Huang L, Hull JJ, Iatrou K, Iga M, Kanost MR, Kotwica J, Li C, Li J, Liu J, Lundmark M, Matsumoto S, Meyering-Vos M, Millichap PJ, Monteiro A, Mrinal N, Niimi T, Nowara D, Ohnishi A, Oostra V, Ozaki K, Papakonstantinou M, Popadic A, Rajam MV, Saenko S, Simpson RM, Soberon M, Strand MR, Tomita S, Toprak U, Wang P, Wee CW, Whyard S, Zhang W, Nagaraju J, Ffrench-Constant RH, Herrero S, Gordon K, Swevers L, Smagghe G (2011) RNA interference in Lepidoptera: an overview of successful and unsuccessful studies and implications for experimental design. J Insect Physiol 57:231–245

The International Silkworm Genome Consortium (2008) The genome of a lepidopteran model insect, the silkworm *Bombyx mori*. Insect Biochem Mol Biol 38:1036–1045

Tomita M, Munetsuna H, Sato T, Adachi T, Hino R, Hayashi M, Shimizu K, Nakamura N, Tamura T, Yoshizato K (2003) Transgenic silkworms produce recombinant human type III procollagen in cocoons. Nat Biotechnol 21:52–56

Uchino K, Imamura M, Shimizu K, Kanda T, Tamura T (2007) Germ line transformation of the silkworm, *Bombyx mori*, using the transposable element *Minos*. Mol Genet Genomics 277:213–220

Uchino K, Sezutsu H, Imamura M, Kobayashi I, Tatematsu K, Iizuka T, Yonemura N, Mita K, Tamura T (2008) Construction of a *piggyBac*-based enhancer trap system for the analysis of gene function in silkworm *Bombyx mori*. Insect Biochem Mol Biol 38:1165–1173

Uhlirova M, Foy BD, Beaty BJ, Olson KE, Riddiford LM, Jindra M (2003) Use of Sindbis virus-mediated RNA interference to demonstrate a conserved role of broad-complex in insect metamorphosis. Proc Natl Acad Sci U S A 100:15607–15612

Urnov FD, Rebar EJ, Holmes MC, Zhang HS, Gregory PD (2010) Genome editing with engineered zinc finger nucleases. Nat Rev Genet 11:636–646

Wang Y, Li Z, Xu J, Zeng B, Ling L, You L, Chen Y, Huang Y, Tan A (2013) The CRISPR/Cas system mediates efficient genome engineering in *Bombyx mori*. Cell Res 23:1414–1416

Watanabe T, Ochiai H, Sakuma T, Horch HW, Hamaguchi N, Nakamura T, Bando T, Ohuchi H, Yamamoto T, Noji S, Mito T (2012) Non-transgenic genome modifications in a hemimetabolous insect using zinc-finger and TAL effector nucleases. Nat Commun 3:1017

Xia Q, Zhou Z, Lu C, Cheng D, Dai F, Li B, Zhao P, Zha X, Cheng T, Chai C, Pan G, Xu J, Liu C, Lin Y, Qian J, Hou Y, Wu Z, Li G, Pan M, Li C, Shen Y, Lan X, Yuan L, Li T, Xu H, Yang G, Wan Y, Zhu Y, Yu M, Shen W, Wu D, Xiang Z, Yu J, Wang J, Li R, Shi J, Li H, Su J, Wang X, Zhang Z, Wu Q, Li J, Zhang Q, Wei N, Sun H, Dong L, Liu D, Zhao S, Zhao X, Meng Q, Lan F, Huang X, Li Y, Fang L, Li D, Sun Y, Yang Z, Huang Y, Xi Y, Qi Q, He D, Huang H, Zhang X, Wang Z, Li W, Cao Y, Yu Y, Yu H, Ye J, Chen H, Zhou Y, Liu B, Ji H, Li S, Ni P, Zhang J, Zhang Y, Zheng H, Mao B, Wang W, Ye C, Wong GK, Yang H (2004) A draft sequence for the genome of the domesticated silkworm (*Bombyx mori*). Science 306:1937–1940

Xiao A, Wang Z, Hu Y, Wu Y, Luo Z, Yang Z, Zu Y, Li W, Huang P, Tong X, Zhu Z, Lin S, Zhang B (2013a) Chromosomal deletions and inversions mediated by TALENs and CRISPR/Cas in zebrafish. Nucleic Acids Res 41:e141

Xiao A, Wu Y, Yang Z, Hu Y, Wang W, Zhang Y, Kong L, Gao G, Zhu Z, Lin S, Zhang B (2013b) EENdb: a database and knowledge base of ZFNs and TALENs for endonuclease engineering. Nucleic Acids Res 41(Database issue):D415–D422

Yamaguchi J, Mizoguchi T, Fujiwara H (2011) siRNAs induce efficient RNAi response in *Bombyx mori* embryos. PLoS One 6:e25469

Yu Z, Ren M, Wang Z, Zhang B, Rong YS, Jiao R, Gao G (2013) Highly efficient genome modifications mediated by CRISPR/Cas9 in *Drosophila*. Genetics 195:289–291

Chapter 6
Genome Editing in Sea Urchin

Naoaki Sakamoto

Abstract The sea urchin has long been used for biological research as a model organism. Methods for gene transfer, gene disruption and analysis of gene expression in the sea urchin embryo have been established, and the gene regulatory network of sea urchin endomesoderm specification has been elucidated recently. To analyze the functional linkage among regulatory genes for construction of the gene regulatory network, the function of each regulatory gene is perturbed, and the subsequent qualitative and quantitative analyses of the responses are determined. For this purpose, gene knockdown by morpholino antisense oligonucleotides has been routinely performed. Furthermore, to investigate the *cis*-regulatory mechanism responsible for spatiotemporal expression of regulatory genes, transfer of an external reporter gene has been carried out. However, genetic engineering, such as knockout and knock-in are available only in particular model organisms and have not been applied to sea urchin research. Recently, a new technology for targeted genome editing using programmable nucleases, such as zinc finger nuclease (ZFN), transcription activator-like effector nuclease (TALEN) and RNA-guided endonucleases CRISPR/Cas9 have been developed. In this chapter, I introduce the successful applications of ZFN and TALEN technologies in sea urchin research.

Keywords Echinoderm • Ets • Gene regulatory network • HesC • Microinjection • Morpholino • Sea urchin

6.1 Introduction on Sea Urchins

The echinoderms are deuterostome invertebrate animals that are in the lineage leading to the vertebrates. The echinoderms are characterized by fivefold radial symmetry of the adult body plan, an endoskeleton that is comprised of small calcium carbonate plates, and the presence of a water vascular system. The echinoderms have five classes: sea urchins, sea stars, sea cucumbers, brittle stars and sea lilies.

N. Sakamoto (✉)
Department of Mathematical and Life Sciences, Graduate School of Science, Hiroshima University, 1-3-1 Kagamiyama, Higashi-Hiroshima 739-8526, Japan
e-mail: naosaka@hiroshima-u.ac.jp

T. Yamamoto (ed.), *Targeted Genome Editing Using Site-Specific Nucleases*,
DOI 10.1007/978-4-431-55227-7_6

There are approximately 1,000 species of sea urchins living in various marine environments. Several species of sea urchins (*Strongylocentrotus*, *Hemicentrotus*, *Lytechinus* and *Paracentrotus*) have been generally used for biological research for the following reasons. Gametes can be obtained easily by coelomic injection of KCl solution. The early embryos of many commonly used sea urchin species are highly transparent; therefore, their internal structures can be observed clearly. The early development of sea urchin embryos is synchronous; therefore, many synchronously developing embryos can be used for biochemical experiments. Methods for gene transfer into sea urchin eggs have been established, and transgenic manipulation has been performed with reporter gene constructs in studies of sea urchin embryogenesis (Rast 2000; Arnone et al. 2004). Furthermore, gene knockdown by morpholino antisense oligonucleotides (MASOs) can be routinely performed in the analysis of gene function (Angerer and Angerer 2004). Using these techniques, the gene regulatory network of sea urchin endomesoderm specification has been elucidated, and the development of sea urchins has been envisioned as the functional interactions among regulatory genes that encode transcription factors (Davidson et al. 2002; Oliveri and Davidson 2004).

The genome of the sea urchin *Strongylocentrotus purpuratus* was sequenced and found to contain 814 million base pairs and was estimated to have 23,300 genes (Sea Urchin Genome Sequencing Consortium 2006). The genome size of the sea urchin is about one fourth that of humans, but the number of genes of the sea urchin is equivalent to that of humans. Furthermore, a reciprocal BLAST search of the genome between sea urchin and other species revealed that the sea urchin genome contains a number of orthologous genes to mammals (Materna et al. 2006). On the other hand, categorization of predicted gene models by their best match to conserved protein domains indicated that among the 50 most abundant gene model groups, 16 are not found among the most numerous of the genomes of mouse, *Ciona intestinalis*, *Drosophila melanogaster* and *Caenorhabditis elegans* (Materna et al. 2006). Furthermore, a significant number of protein domains that match proteins from other genomes were not found in the sea urchin genome (Materna et al. 2006). Taken together, the sea urchin genome shares numerous genes with other animals, but there are some differences that generate the unique properties of sea urchins.

The sea urchin genome is highly polymorphic relative to the genome of other species. In *S. purpuratus*, the differences among single copy DNA sequences between individuals was estimated as about 4 % (Britten et al. 1978). The assembly of the *S. purpuratus* genome revealed that there is at least one single nucleotide polymorphism (SNP) per 100 bases, as well as a comparable frequency of indel variants (Sea Urchin Genome Sequencing Consortium 2006). Furthermore, in *Hemicentrotus pulcherrimus*, SNPs were detected not only in protein coding sequences, but also in noncoding regions, such as enhancers and untranslated regions (Yamamoto et al. 2007).

In sea urchin research, culture and inbreeding techniques have been established. In *S. purpuratus*, inbred lines were constructed by successive sibling matings, and an inbred line of more than seven-generations of sibling matings has been produced (Leahy et al. 1994). Progeny of crosses among F_2 and F_3 siblings of this line showed normal embryogenesis; however, they showed lethality, morphological malformation and arrest at high frequency during postembryonic larval development (Leahy et al. 1994), probably because of the highly polymorphic feature of the sea urchin genome. However, the established inbred line is healthy and fertile, and inbreeding increased the homozygosity of the genome (Cameron et al. 1999).

6.2 Disruption of Gene Functions

Sea urchins have been used to investigate the molecular mechanisms of early development. A number of genes that are involved in morphogenesis during early development have been identified. From the analyses of their *cis*-regulatory elements and the function of each gene, functional interactions among regulatory genes have been clarified and a gene regulatory network (GRN) has been constructed (Davidson et al. 2002; Oliveri and Davidson 2004). To construct the GRN, the expression and/or the function of each component must be perturbed, and the responses to the perturbation should be analyzed qualitatively and quantitatively.

The perturbation of each regulatory gene has been achieved by ectopic overexpression of the mRNA or by inhibition of gene expression and function. Although a pharmacological perturbation by an inhibitor could be used to inhibit the activity of the gene product, the toxicity and the side effects have to be considered carefully. Besides pharmacological perturbation, microinjection of a purified monoclonal antibody (Morris and Scholey 1997; Kanungo et al. 1999) or mRNA encoding a single chain antibody (Bogarad et al. 1998) have been used to disturb the function of a gene product. Moreover, microinjection of mRNA encoding a dominant negative polypeptide, which is either a truncated polypeptide (Kurokawa et al. 1999) or a polypeptide fused to the Engrailed repressor domain (Oliveri et al. 2002), has also been used.

For perturbation of gene expressions, two types of approaches, RNA interference (RNAi) and MASOs, are generally used. Although the expression of genes required for the RNAi pathway during development has been reported in *S. purpuratus* (Song and Wessel 2007) and the involvement of Dicer in the early development of the sea urchin has been demonstrated (Okamitsu et al. 2010), an experimental approach to block gene expression by siRNA was unsuccessful. This was probably because the RNA molecules targeting mRNAs are unstable and cannot be retained at high concentration after injection (Smith and Davidson 2008). Therefore, MASO has been used routinely for the perturbation of gene function in sea urchin research (Angerer and Angerer 2004).

A MASO is a DNA analog in which the riboside moiety of each subunit is converted to a morpholine moiety and a phosphorodiamidate is used for intersubunit linkage instead of a phosphodiester (Summerton and Weller 1997). DNA and S-DNA antisense oligonucleotides can block translation predominantly by RNaseH-cleavage of duplex of mRNA and antisense oligonucleotides. On the other hand, MASO is resistant to enzymatic digestion by cellular nucleases and is less toxic;

it physically blocks translation only by interfering with the assembly of the ribosome onto the target mRNA. Therefore, MASOs should be designed to be complementary to the 5′ leader sequences or to the first 25 nucleotides 3′ to the translation start site (Summerton 1999; Heasman 2002). Furthermore, MASOs targeted to exon–intron junction sites in primary transcripts can interfere with RNA splicing, resulting in the production of altered gene products.

Recently, targeted genome editing with programmable nucleases has been reported in animals, plants and cultured cells. Three types of programmable nucleases, zinc finger nuclease (ZFN), transcription activator-like effector nuclease (TALEN) and RNA-guided endonuclease CRISPR/Cas9 have been used for this purpose (Carroll 2014; Kim and Kim 2014). The engineered nucleases, ZFN and TALEN, are comprised of a customized DNA-binding domain and a nuclease domain of the restriction enzyme FokI. A pair of engineered nucleases binds to their target site within the genome in a sequence-specific manner and induce a double-strand break (DSB). On the other hand, the RNA-guided endonuclease CRISPR/Cas9 uses short guide RNA (gRNA) to take the Cas9 nuclease to the target site and induce a DSB. Either nonhomologous end joining (NHEJ) or homology-directed repair (HDR) pathways are used to repair the induced DSBs. NHEJ is an error-prone process of repair; therefore, insertions and/or deletions can be induced in the target site during the NHEJ repair process.

Targeted genome editing using the engineered nuclease in sea urchin embryos was first reported by Ochiai et al. (2010). They employed ZFNs in which zinc fingers were used as the DNA-binding domain. ZFNs with a three-finger array were used for gene disruption in the sea urchin, *Hemicentrotus pulcherrimus*. Since wild populations of sea urchins are generally used for the research purposes and the sea urchin genome is highly polymorphic, DNA variations within the genomic region to be targeted were analyzed in the genomic DNA extracted from several individuals. Then, to generate functional ZFNs, a ZF randomized library was constructed, and each ZF module that recognize three base pairs of the 9 bp target sequence was selected by bacterial one-hybrid screening. Zinc-finger arrays show context-dependence between adjacent zinc-fingers; therefore, a second round bacterial one-hybrid screening was performed after conjunction of the selected three fingers. Furthermore, the site-specific cleavage activity of ZFN with the selected ZF array was evaluated by a single-strand annealing assay (SSA assay).

In the first application of ZFN to the sea urchin, the sea urchin *HesC* gene, which is expressed in the whole embryonic region except for the micromere lineage during early development, was selected as the target (Ochiai et al. 2010). HesC represses micromere-specific regulatory genes (Revilla-i-Domingo et al. 2007); therefore, it is expected that disruption of HesC would cause all of the embryonic cells to express micromere-specific genes. A pair of ZFNs targeting the third exon of *HpHesC* gene, which is the *H. pulcherrimus* homolog of *HesC*, was constructed by fusion of the selected ZF array and a nuclease domain of FokI. In this experiment, the FokI nuclease variants RR and DD were employed for each ZFN, respectively, to avoid the formation of homodimers of each ZFN (Szczepek et al. 2007).

To introduce the ZFNs into sea urchin embryos, mRNAs encoding HpHesC ZFNs were synthesized in vitro and microinjected into the fertilized eggs of sea urchins. An increase in the number of primary mesenchyme cells (PMCs) within the blastocoel was observed in approximately 10 % of embryos injected with a pair of HpHesC ZFN mRNAs. This phenotype is consistent with that of HpHesC-knockdown by MASO. Analyses of mutations in the genomic DNA extracted from ZFN mRNA-injected embryos revealed that introduction of deletions and insertions started by 4 hours postfertilization (hpf) (eight-cell stage) and the number of mutations reached the maximum level at 12 hpf (unhatched blastula stage). Furthermore, in the ZFN mRNA-injected *H. pulcherrimus* embryos, 44 % of *HesC* genes were disrupted by frameshift mutations at 8 hpf (morula stage). Thus, despite the high frequency of frameshift mutations, a relatively low frequency of the affected phenotype was observed. This is probably because only a single allele of *HesC* gene was disrupted in many cells. Alternatively, since the expression of *HesC* gene is initiated at the morula stage (8 hpf) in *S. purpuratus* (Revilla-i-Domingo et al. 2007), significant mutagenesis could have occurred after the zygotic expression of the *HesC* gene. Further effective methods for genome editing may be required to achieve an efficient knockout of the target genes.

Targeted genome editing with another type of the engineered nuclease, TALENs, was recently reported by Hosoi et al. (2014). TALENs contain the TAL effector from *Xanthomonas sp.* as the DNA-binding domain. TALENs were constructed that targeted the third exon of *HpEts* gene, the *H. pulcherrimus* homolog of *Ets1*, which plays essential role in specification of PMCs in sea urchin embryos (Kurokawa et al. 1999). The Golden Gate cloning method was used to assemble the TALE repeats (Cermak et al. 2011; Sakuma et al. 2013a). Furthermore, to increase the efficiency of gene disruption, they employed Platinum TALEN, which contains periodically patterned repeat variants harboring non-repeat-variable di-residue (non-RVD) variations (Sakuma et al. 2013b).

To introduce the TALENs into sea urchin embryos, mRNA encoding HpEts TALENs were synthesized in vitro and microinjected into fertilized sea urchin eggs. When a pair of HpEts TALENs was introduced, 12.6 % of HpEts TALEN pair-injected embryos showed impairment of skeletogenesis, such as loss or incomplete formation of skeletal spicules. This phenotype is consistent with that observed by microinjection of mRNA encoding a dominant-negative form of HpEts (Kurokawa et al. 1999). However, 2.9 % of embryos microinjected with a single HpEts TALEN showed impairment of skeletogenesis, probably because of the use of a homodimer type of FokI nuclease domain. Restriction fragment length polymorphism (RFLP) analysis of the target site, which was extracted and amplified from HpEts TALEN pair-injected embryos, indicated that obvious mutagenesis was first detected at 8 hpf (morula stage), and the amount of mutations increased at 12 hpf (unhatched blastula stage) and reached a plateau at approximately 24 hpf (mesenchyme blastula stage). Compared with the analysis of ZFN mRNA-introduced mutations (Ochiai et al. 2010), it seems that the timing of TALEN-mediated mutagenesis might be later than ZFN-mediated mutagenesis, although they cannot be simply compared because of the difference in their target sites. Furthermore, sequencing analysis of mutations introduced into the *HpEts* gene by the TALEN pair indicated that 51.9 % of *HpEts* genes were disrupted by frameshift mutations at 24 hpf.

Since the zygotic expression of *HpEts* gene starts at the cleavage stage and reaches maximum at the unhatched blastula stage (Yajima et al. 2010), the explanation of the relatively low frequency of the disrupted phenotype may be the same as described above for HesC ZFN. Moreover, the *HpEts* gene is expressed only in the micromere lineage; therefore, mutagenesis must occur on both alleles in this lineage to obtain the disrupted phenotype. Therefore, the mosaic pattern of the mutagenesis may be related to the low frequency of the disrupted phenotype, and reduction of mosaicism may be required for efficient genome editing in the sea urchin.

6.3 Use of Reporters to Visualize Gene Expression

Sea urchin embryos have been used to study *cis*-regulatory elements that are responsible for spatial and temporal expression of genes. To clarify the mechanism that controls the transcription of a particular gene, transgenic manipulations have been carried out (Rast 2000; Arnone et al. 2004). For this purpose, a large upstream regulatory region of the gene of interest was fused to a reporter gene, such as the luciferase gene. These exogenous fusion gene constructs have been shown to be expressed in sea urchin embryos similarly to the endogenous gene. However, *cis-regulatory* elements are dispersed within a gene, being located not only upstream of the transcription start point, but also downstream, such as in introns. Therefore, analyses of *cis*-regulatory elements using a reporter gene should be performed in the long genomic context. Furthermore, the compositional properties of the DNA surrounding a *cis*-regulatory region of a gene can affect *cis*-regulatory control of gene expression in the sea urchin (Mahmud et al. 2008). A large bacterial artificial chromosome (BAC) clone containing a full set of *cis*-regulatory elements within the natural genomic context was constructed and used for *cis*-regulatory analysis (Smith 2008). At present, BAC libraries of several sea urchin species, in which each clone contains >100 bp sequences, are available. To generate a BAC reporter construct, a cassette of the green fluorescent protein (GFP) coding region is knocked in in-frame into an exon by homologous recombination in *Escherichia coli* (Yu et al. 2000; Lee et al. 2001).

Exogenous DNA constructs can be introduced into sea urchin eggs by microinjection or particle gun methods (Akasaka et al. 1995; Cheers and Ettensohn 2004). The injected DNA that has been linearized by a restriction endonuclease undergoes random end-to-end concatenation rapidly in the egg cytoplasm and is then amplified by DNA replication (McMahon et al. 1985). On the other hand, supercoiled circular plasmids microinjected into sea urchin eggs are retained with the embryo until the pluteus stage, but they are neither concatenated nor amplified within the eggs (McMahon et al. 1985). The concatenated plasmids are subsequently integrated into the sea urchin genome in a mosaic fashion during cleavage, resulting in a mosaic pattern of expression of the exogenous genes (Flytzanis et al. 1985; Hough-Evans et al. 1988). Utilization of *I-SceI* meganuclease-mediated transgenesis was shown to reduce the mosaicism of the transgene expression in the sea urchin (Ochiai et al. 2008). To increase the transgene expression, carrier DNA, which is prepared by enzymatic digestion of sea urchin sperm DNA, is used. When transgenes were separated by more than 3.5-fold excess of carrier DNA, they functioned more efficiently, probably because of transcriptional interference in the absence of the carrier (Franks et al. 1990). The *Ars* insulator can protect transgenes from silencing by the surrounding repressive chromatin environment (Akasaka et al. 1999; Yajima et al. 2007). Furthermore, the *Ars* insulator increases the frequency of transgene expression, reduces the ectopic expression level and promotes the integration of transgenes into the genome (Ochiai et al. 2007).

Using the ZFN technology, targeted insertion of a reporter gene has also been successful (Ochiai et al. 2012). The target gene was the *HpEts* gene, which encodes a transcription factor responsible for PMC differentiation and is expressed in the presumptive PMCs at the hatched blastula stage and in PMCs at the mesenchyme blastula stage (Kurokawa et al. 1999). A pair of ZFNs targeting the region upstream of the stop codon in the *HpEts* gene was generated by bacterial one-hybrid (B1H) and SSA screenings. The targeting donor construct for HDR was comprised of $\approx$1 kb of homology arms and a histone H2B-GFP fusion gene. In this donor construct, a histone *H2B-GFP* fusion gene was bicistronically linked to the *HpEts* gene via the 'self- cleaving' 2A peptide sequence to avoid disturbances in the expression and function of the *HpEts* gene (Szymczak et al. 2004). To increase the efficiency of the ZFN-mediated targeted insertion, HpEts ZFN target sequences were inserted at both ends of the donor cassette, so that the linear targeting donor cassette is excised from the circular plasmid within the injected embryos. In *Drosophila*, high frequency ZFN-mediated targeted gene modification was achieved by production of a linear donor in situ (Beumer et al. 2006). Furthermore, to improve the efficiency of HDR, mRNA encoding the carboxy-terminal tandem BRCT repeat of DNA ligase IV (*DN-lig4*) was co-injected. Overexpression of this mRNA was expected to induce a dominant-negative effect and subsequently increase the frequency of HDR by reducing the occurrence of NHEJ (Wu et al. 2009). Eventually, targeted insertion of *H2B-GFP* gene into the *HpEts* gene was successful in the sea urchin, and in situ linearization of the donor vector improved the efficiency (Ochiai et al. 2012). Co-injection of *DN-lig4* mRNA resulted in a slight increase in efficiency. There are at lease two NHEJ pathways in some cell types, and DNA ligase IV seems to be required for only one of them (Beumer et al. 2006); therefore, the effect of disturbance of DNA ligase IV might be weak. Using this knock-in embryo, the fluorescence intensity of each single PMC in living embryos during development was measured to monitor the expression of the *HpEts* gene, and variation in *HpEts* gene expression among the PMCs was demonstrated quantitatively. Thus, engineered nuclease-mediated targeted transgene insertion is feasible in the sea urchin and may be applied to various quantitative analyses of gene expression.

6.4 Application of Genome Editing in Sea Urchin Research

Gene disruption studies using MASOs has been performed in the functional analyses of regulatory genes during embryogenesis, and genome editing technology is now available in sea urchin research. For the analysis of the early embryogenesis of the sea urchin, the disruption of target genes on both alleles should occur at an early period of embryogenesis. Therefore, earlier expression and higher efficiency of the engineered nucleases may be required to apply these technologies to the analysis of early embryogenesis. Furthermore, the mutations induced by the engineered nucleases are heterogeneous, and if this technology could be combined with the establishment of inbreeding lines of sea urchins, the production of knockout lines of sea urchins with homogeneous mutations may be possible.

Once a mutation is introduced in the genome by the genome editing, it is transmitted to the descendants. Therefore, this technology could enable us to investigate the functions of gene products during the larval stage and the development of adult body plan within the adult rudiment. Live cell imaging may also be used to monitor the gene expressions within the adult rudiment.

References

Akasaka K, Nishimura A, Hijikata K, Iuchi Y, Morokuma J, Takahashi M, Morikawa H, Shimada H (1995) Introduction of DNA into sea urchin eggs by particle gun. Mol Mar Biol Biotechnol 4:255–261

Akasaka K, Nishimura A, Takata K, Mitsunaga K, Mibuka F, Ueda H, Hirose S, Tsutsui K, Shimada H (1999) Upstream element of the sea urchin arylsulfatase gene serves as an insulator. Cell Mol Biol (Noisy-le-grand) 45:555–565

Angerer LM, Angerer RC (2004) Disruption of gene function using antisense morpholinos. Methods Cell Biol 74:699–711

Arnone MI, Dmochowski IJ, Gache C (2004) Using reporter genes to study *cis*-regulatory elements. Methods Cell Biol 74:621–652

Beumer K, Bhattacharyya G, Bibikova M, Trautman JK, Carroll D (2006) Efficient gene targeting in *Drosophila* with zinc-finger nucleases. Genetics 172:2391–2403

Bogarad LD, Arnone MI, Chang C, Davidson EH (1998) Interference with gene regulation in living sea urchin embryos: transcription factor knock out (TKO), a genetically controlled vector for blockade of specific transcription factors. Proc Natl Acad Sci U S A 95:14827–14832

Britten RJ, Cetta A, Davidson EH (1978) The single-copy DNA sequence polymorphism of the sea urchin *Strongylocentrotus purpuratus*. Cell 15:1175–1186

Cameron RA, Leahy PS, Britten RJ, Davidson EH (1999) Microsatellite loci in wild-type and inbred *Strongylocentrotus purpuratus*. Dev Biol 208:255–264

Carroll D (2014) Genome engineering with targetable nucleases. Annu Rev Biochem 83:14.1–14.31

Cermak T, Doyle EL, Christian M, Wang L, Zhang Y, Schmidt C, Baller JA, Somia NV, Bogdanove AJ, Voytas DF (2011) Efficient design and assembly of custom TALEN and other TAL effector-based constructs for DNA targeting. Nucleic Acids Res 39:e82

Cheers MS, Ettensohn CA (2004) Rapid microinjection of fertilized eggs. Methods Cell Biol 74:287–310

Davidson EH, Rast JP, Oliveri P, Ransick A, Calestani C, Yuh CH, Minokawa T, Amore G, Hinman V, Arenas-Mena C, Otim O, Brown CT, Livi CB, Lee PY, Revilla R, Rust AG, Zj P, Schilstra

MJ, Clarke PJ, Arnone MI, Rowen L, Cameron RA, McClay DR, Hood L, Bolouri H (2002) A genomic regulatory network for development. Science 295:1669–1678

Flytzanis CN, McMahon AP, Hough-Evans BR, Katula KS, Britten RJ, Davidson EH (1985) Persistence and integration of cloned DNA in postembryonic sea urchins. Dev Biol 108:431–442

Franks RR, Anderson R, Moore JG, Hough-Evans BR, Britten RJ, Davidson EH (1990) Competitive titration in living sea urchin embryos of regulatory factors required for expression of the CyIIIa actin gene. Development 110:31–40

Heasman J (2002) Morpholino oligos: making sense of antisense? Dev Biol 243:209–214

Hosoi S, Sakuma T, Sakamoto N, Yamamoto T (2014) Targeted mutagenesis in sea urchin embryos using TALENs. Dev Growth Differ 56:92–97

Hough-Evans BR, Britten RJ, Davidson EH (1988) Mosaic incorporation and regulated expression of an exogenous gene in the sea urchin embryo. Dev Biol 129:198–208

Kanungo J, Empson RM, Rasmussen H (1999) Microinjection of an antibody to the Ku protein arrests development in sea urchin embryos. Biol Bull 197:341–347

Kim H, Kim JS (2014) A guide to genome engineering with programmable nucleases. Nat Rev Genet 15:321–334

Kurokawa D, Kitajima T, Mitsunaga-Nakatsubo K, Amemiya S, Shimada H, Akasaka K (1999) HpEts, an ets-related transcription factor implicated in primary mesenchyme cell differentiation in the sea urchin embryo. Mech Dev 80:41–52

Leahy PS, Cameron RA, Knox MA, Britten RJ, Davidson EH (1994) Development of sibling inbred sea urchins: normal embryogenesis, but frequent postembryonic malformation, arrest and lethality. Mech Dev 45:255–268

Lee EC, Yu D, Martinez de Velasco J, Tessarollo L, Swing DA, Court DL, Jenkins NA, Copeland NG (2001) A highly efficient *Escherichia coli*-based chromosome engineering system adapted for recombinogenic targeting and subcloning of BAC DNA. Genomics 73:56–65

Mahmud AA, Amore G, Bernardi G (2008) Compositional genome contexts affect gene expression control in sea urchin embryo. PLoS One 3:e4025

Materna SC, Berney K, Cameron RA (2006) The *S. purpuratus* genome: a comparative perspective. Dev Biol 300:485–495

McMahon AP, Flytzanis CN, Hough-Evans BR, Katula KS, Britten RJ, Davidson EH (1985) Introduction of cloned DNA into sea urchin egg cytoplasm: replication and persistence during embryogenesis. Dev Biol 108:420–430

Morris RL, Scholey JM (1997) Heterotrimeric kinesin-II is required for the assembly of motile 9+2 ciliary axonemes on sea urchin embryos. J Cell Biol 138:1009–1022

Ochiai H, Sakamoto N, Suzuki K, Akasaka K, Yamamoto T (2008) The *Ars* insulator facilitates *I-SceI* meganuclease-mediated transgenesis in the sea urchin embryo. Dev Dyn 237: 2475–2482

Ochiai H, Fujita K, Suzuki K, Nishikawa M, Shibata T, Sakamoto N, Yamamoto T (2010) Targeted mutagenesis in the sea urchin embryo using zinc-finger nucleases. Genes Cells 15:875–885

Ochiai H, Sakamoto N, Fujita K, Nishikawa M, Suzuki KI, Matsuura S, Miyamoto T, Sakuma T, Shibata T, Yamamoto T (2012) Zinc-finger nuclease-mediated targeted insertion of reporter genes for quantitative imaging of gene expression in sea urchin embryos. Proc Natl Acad Sci U S A 109:10915–10920

Okamitsu Y, Yamamoto T, Fujii T, Ochiai H, Sakamoto N (2010) Dicer is required for the normal development of sea urchin, *Hemicentrotus pulcherrimus*. Zoolog Sci 27:477–486

Oliveri P, Davidson EH (2004) Gene regulatory network controlling embryonic specification in the sea urchin. Curr Opin Genet Dev 14:351–360

Oliveri P, Carrick DM, Davidson EH (2002) A regulatory gene network that directs micromere specification in the sea urchin embryo. Dev Biol 246:209–228

Rast JP (2000) Transgenic manipulation of the sea urchin embryo. Methods Mol Biol 136: 365–373

Revilla-i-Domingo R, Oliveri P, Davidson EH (2007) A missing link in the sea urchin embryo gene regulatory network: *hesC* and the double-negative specification of micromeres. Proc Natl Acad Sci U S A 104:12383–12388

Sakuma T, Hosoi S, Woltjen K, Suzuki KI, Kashiwagi K, Wada H, Ochiai H, Miyamoto T, Kawai N, Sasakura Y, Matsuura S, Okada Y, Kawahara A, Hayashi S, Yamamoto T (2013a) Efficient TALEN construction and evaluation methods for human cell and animal applications. Genes Cells 18:315–326

Sakuma T, Ochiai H, Kaneko T, Mashimo T, Tokumasu D, Sakane Y, Suzuki K, Miyamoto T, Sakamoto N, Matsuura S, Yamamoto T (2013b) Repeating pattern of non-RVD variations in DNA-binding modules enhances TALEN activity. Sci Rep 3:3379

Sea Urchin Genome Sequencing Consortium (2006) The genome of the sea urchin *Strongylocentrotus purpuratus*. Science 314:941–952

Smith J (2008) A protocol describing the principles of *cis*-regulatory analysis in the sea urchin. Nat Protoc 3:710–718

Smith J, Davidson EH (2008) A new method, using *cis*-regulatory control, for blocking embryonic gene expression. Dev Biol 318:360–365

Song JL, Wessel GM (2007) Genes involved in the RNA interference pathway are differentially expressed during sea urchin development. Dev Dyn 236:3180–3190

Summerton J (1999) Morpholino antisense oligomers: the case for an RNase H-independent structural type. Biochim Biophys Acta 1489:141–158

Summerton J, Weller D (1997) Morpholino antisense oligomers: design, preparation, and properties. Antisense Nucleic Acid Drug Dev 7:187–195

Szczepek M, Brondani V, Büchel J, Serrano L, Segal DJ, Cathomen T (2007) Structure-based redesign of the dimerization interface reduces the toxicity of zinc-finger nucleases. Nat Biotechnol 25:786–793

Szymczak AL, Workman CJ, Wang Y, Vignali KM, Dilioglou S, Vanin EF, Vignali DA (2004) Correction of multi-gene deficiency in vivo using a single 'self-cleaving' 2A peptide-based retroviral vector. Nat Biotechnol 22:589–594

Wu PY, Frit P, Meesala S, Dauvillier S, Modesti M, Andres SN, Huang Y, Sekiguchi J, Calsou P, Salles B, Junop MS (2009) Structural and functional interaction between the human DNA repair proteins DNA ligase IV and XRCC4. Mol Cell Biol 29:3163–3172

Yajima M, Kiyomoto M, Akasaka K (2007) *Ars* insulator protects transgenes from long-term silencing in sea urchin larva. Dev Genes Evol 217:331–336

Yajima M, Umeda R, Fuchikami T, Kataoka M, Sakamoto N, Yamamoto T, Akasaka K (2010) Implication of *HpEts* in gene regulatory networks responsible for specification of sea urchin skeletogenic primary mesenchyme cells. Zoolog Sci 27:638–646

Yamamoto T, Kawamoto R, Fujii T, Sakamoto N, Shibata T (2007) DNA variations within the sea urchin *Otx* gene enhancer. FEBS Lett 581:5234–5240

Yu D, Ellis HM, Lee EC, Jenkins NA, Copeland NG, Court DL (2000) An efficient recombination system for chromosome engineering in *Escherichia coli*. Proc Natl Acad Sci U S A 97: 5978–5983

Chapter 7
Genome Editing in Ascidians

Nicholas Treen and Yasunori Sasakura

Abstract Genome editing has the potential to provide new approaches to investigate biological questions, and can provide simpler experimental approaches than more established techniques. Tunicates are a group of diverse marine organisms found in seas throughout the world. Tunicates, and in particular, ascidians are attractive model organisms, particularly for the experimental study of early animal development due to the relative simplicity of development, and the wide range of experimental techniques available, particularly for *Ciona intestinalis*. In this chapter we will discuss the current state of techniques for performing gene knockdowns or knockouts in ascidians and describe the recent progress in using recently developed genome editing technologies utilizing custom nucleases. These new experimental approaches are particularly suited to investigating the later stages of ascidian development, after the tailbud embryo has formed, and have the potential to open up exciting new opportunities to understand the unknown processes of animal development.

Keywords Ascidians • Ciona • CRISPR/Cas9 • Genome editing • TALENs • Zinc finger nuclease

7.1 Introduction: Experimental Techniques for Tunicates

7.1.1 Non-Ascidian Tunicates

Ascidians are a class of exclusively marine filter, sessile, filter-feeding animals. Ascidians are grouped along with larvaceans and thaliaceans as tunicates. The major characteristic of tunicates is the presence of a cellulose tunic that is used to cover and protect major regions of both the larval and adult body (Lemaire 2011). Larvaceans (Fig. 7.1a) are partially interesting due to their exceptionally short life

N. Treen (✉) • Y. Sasakura
Shimoda Marine Research Center, University of Tsukuba,
Shimoda, Shizuoka 415-0025, Japan
e-mail: ntreen@kurofune.shimoda.tsukuba.ac.jp

T. Yamamoto (ed.), *Targeted Genome Editing Using Site-Specific Nucleases*,
DOI 10.1007/978-4-431-55227-7_7

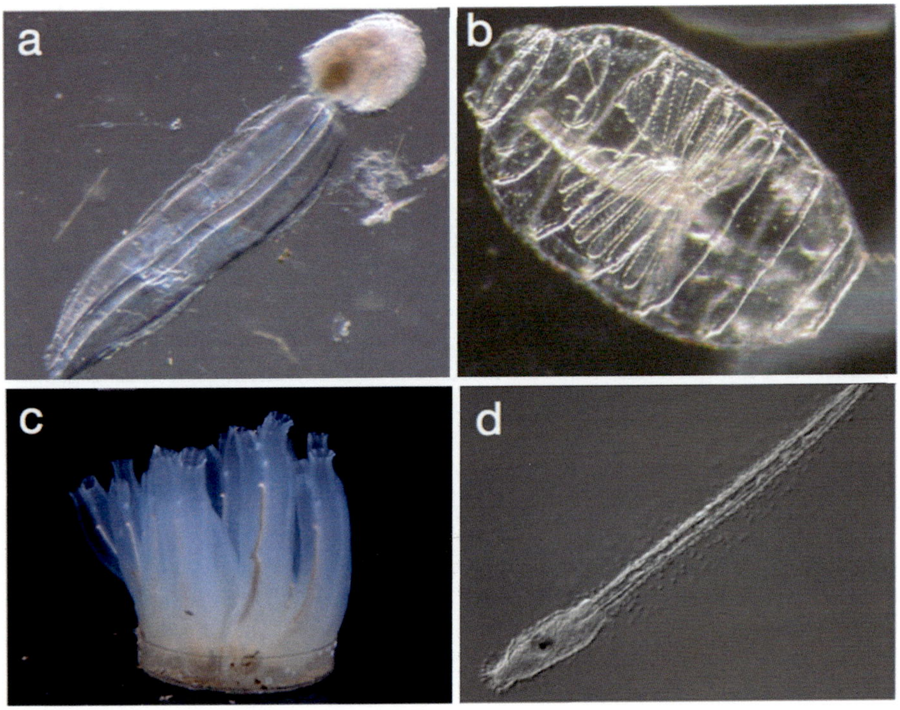

Fig. 7.1 Tunicates: (**a**) An adult larvacean, the house that typically covers and protects the body has been removed. (**b**) An individual thaliacean. (**c**) A plastic petri dish with approximately 10 healthy gravid *Ciona intestinalis* adults attached. Inside the body eggs (*orange/brown color*) and sperm (*white color*) can be easily seen. (**d**) A single *ciona intestinalis* larvae at the swimming larval stage showing normal development. The notochord has fully formed and can be seen throughout the length of the tail. Note the faint outline around the body containing the cellulose tunic and maternally derived test cells

span and compact genomes. *Oikopleura dioica* has been studied in the most detail; with a minimum generation time of 2 days and a genome of 51mega bases (Mb) that has an exceptionally high gene density and an unprecedented reduction of introns (Seo et al. 2001). Thaliaceans (Fig. 7.1b) form large pelagic colonies of multiple individuals (Alldredge and Madin 1982). They have a poorly described viviparous life cycle that may not easily lend itself to laboratory analysis. In both these animal orders, established methods of introducing exogenous genes have not been reliably established making their current potential for genome editing low. But if a method could be established for *Oikopleura*, perhaps similar to a recently devised RNAi protocol (Omotezako et al. 2013), the subsequent experimental data obtainable could be exceptional. The potential for genome editing, and other forms of genetic experiments in thaliaceans is currently extremely poor.

7.1.2 Ascidians

Ascidians exist as solitary or colonial species; the main discussion in this chapter will be on solitary ascidians. But there is splendid work on development and allorecognition using colonial ascidians (for example: Kawamura et al. 2012; Voskoboynik et al. 2013a) and with the recent publication of a draft genome sequence for *Botryllus schlosseri* (Voskoboynik et al. 2013b), developing genome-editing tools for colonial ascidians is likely to be a worthwhile endeavor.

For the rest of this chapter we will discuss the potential for genome editing in solitary ascidians, these tunicates that have been used the most extensively for experimental work. Ascidians have a distinctive vase-like appearance with an incurrent and an excurrent syphon (Fig. 7.1c). Ascidians are typically attached to rocks or other substrates in the wild. There are 2,874 described ascidian species (Appeltans et al. 2012). The presence of a notochord and a dorsal neural tube is the main evidence used to classify ascidians (and other tunicates) as chordates (Fig. 7.1d). For much of modern history they were thought to be basal chordates, a claim strengthened by 18 s rDNA sequencing (Wada and Satoh 1994). But a more in-depth study reversed this paradigm showing that tunicates share a more recent common ancestry to craniates (Delsuc et al. 2006). But regardless of these phylogenetic positions, ascidians have proven themselves to be excellent model organisms and ones that have potential to benefit greatly from the new genome targeting methods that are rapidly being developed.

While ascidians have been recorded as animals since antiquity and were involved in some pioneering embryological research in the late nineteenth and early twentieth centuries, modern ascidian developmental biology can be considered to start in 1973. Relying on theoretical advancements on how gene expression and embryonic development are linked (Davidson and Britten 1971), J.R. Whittaker demonstrated the existence of cytoplasmic components that result in defined embryonic states in earlier precursor cells (Whittaker 1973). But the major advancements were made at the Laboratory of Molecular Developmental Biology at Kyoto University from the late twentieth century up to the present day. Of the many advances made were the verification of cell lineages (Nishida and Satoh 1983), and improving the understanding of the regulatory programs being executed within these lineages (Satoh et al. 1996). Culminating in the description of the regulatory networks that occur during ascidian development (Imai et al. 2006). A number of the reviews of ascidian developmental biology have been written. The most in depth of these are two books by N. Satoh covering most the experimental literature of the pre (Satoh 1994) and post (Satoh 2014) molecular eras.

The choice of ascidian species used for experimentation is based on what is easily available to the researcher and amenability to experimental techniques. *Ciona intestinalis* meets both these criteria as it has a wide geographic distribution and is easily available to the majority of researchers who are interested. Modern ascidian research is also performed using *Ciona savignyi*, a species closely related to *C. intestinalis* that can be found in pacific coasts. *Phallusia mammillata* is also used by European researchers as it is easily available in Atlantic and Mediterranean seas, and has an abundant number of eggs with a distinctive optical clarity providing a high potential for future imaging experiments (Robin et al. 2011), although advances in light microscopy may make this less appealing in the future than it was in the past (Amat and Keller 2013). In Japan the edible ascidian *Halocynthia roretzi* was extensively used before work sequencing the *Ciona* genome started. *H. roretzi* embryos are large, making them more convenient for classical embryological work. And while research continues using this organism, introducing DNA or RNA into *H. roretzi* embryos can be challenging.

An essential initial requirement for genome editing is knowledge of the sequence that one wishes to edit. As mentioned above the *C. intestinails* genome has been sequenced and has been extensively annotated (Dehal et al. 2002). The *C. savignyi* genome has also been sequenced (Small et al. 2007). A *H. roretzi* genome-sequencing consortium exists and has established a precedent for releasing their private sequencing data (Kumano et al. 2014). We are anecdotally aware of a number of ascidian genome sequencing projects currently said to be underway, but considering that *C. intestinalis* was the sixth animal to be sequenced, it does not appear that the genome sequencing of additional ascidian species is a high priority, despite the acknowledgement that the similarity between ascidian early development may not be as straightforward as might be typically assumed (Hudson and Yasuo 2008).

For the remainder of this chapter we will concern ourselves with genome editing in *C. intestinalis*. However there is no reason to suggest that genome editing would not work in other ascidian species, the wide range of organisms that are described in this volume indicates that if attempts were made they would probably be successful.

7.2 Experimental Techniques in *Ciona*

The modern repertoire of experimental techniques for *Ciona* is extensive and, with the exception of genome editing, has been thoroughly reviewed elsewhere (Christiaen et al. 2009; Veeman et al. 2011; Stolfi and Christiaen 2012; Sasakura et al. 2012). Only the methods directly related to genome editing will be described here. An essential requirement for genome editing is the ability to introduce the nuclease to the cells of interest. In *Ciona* two methods have been developed that are reliable and accessible to researchers on a daily basis. These are microinjection and electroporation. Microinjection (Fig. 7.2a) is used to introduce DNA, typically a plasmid construct containing a defined *cis*-regulatory element driving the expression of a reporter gene or the misexpression of an endogenous gene. Microinjections can also introduce RNAs that could be used for the overexpression or rescue of specific genes by introducing in-vitro synthesized mRNA, or morpholino antisense oligonucleotides (MASOs) that can bind to RNAs to disrupt their splicing or translation and interfere with the expression of that genes product and any other genes that rely on its faithful expression. Microinjection will typically be performed on unfertilized eggs that will be fertilized after a clutch (typically 20–100) has been injected. The injection process

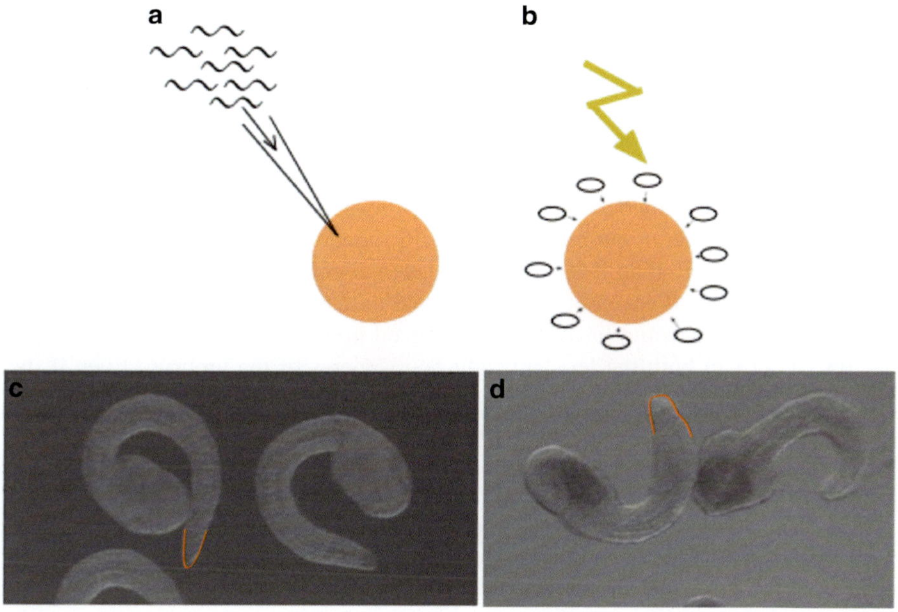

Fig. 7.2 TALEN knockouts in *ciona* embryos. (**a**) Schematic illustrating the introduction of mRNA (*wavy lines*) into eggs by microinjection. (**b**) Schematic illustrating the introduction of plasmid DNA (*small circles*) into eggs by electroporation. (**c**) The result of a control electroporation introducing TALENs not targeting an essential genomic loci for development, (**d**) The result of an electroporation introducing TALENs targeting *Hox12* expressed in the epidermis. In both micrographs note the shape of the tail tips (highlighted with *orange lines*)

does not involve the piercing of the egg such as would be done in the injection of sea urchins, fish or frogs but relies on a localized disturbance caused from negative air pressure after the microinjection needle tip has been pushed against the egg cortex. Individual blastomeres can also be injected allowing lineage specific knockouts, but this is dependent on the shape and position of the cell and the skill of the investigator and becomes more challenging as embryonic development proceeds.

Electroporation (Fig. 7.2b) allows the simultaneous introduction of DNA or RNA into hundreds of 1-cell embryos in a single procedure (Corbo et al. 1997). Around 300 embryos can be typically transformed in a single electroporation. The optimal time to perform electroporations can vary depending on a number of factors such the season and the condition of the eggs, but 15–25 min after fertilization will usually work well. Since a single electroporation requires about 1 minute of hands on time this method can produce around 3,000 synchronously developing embryos in within one hour (including time to prepare the eggs for the procedure). The main disadvantage to this procedure is that a large amount of DNA needed. A single electroporation will require 20–100 μg of purified plasmid DNA. We typically use 60–80 μg to perform a single electroporation. The electroporation of RNA, for example in-vitro synthesized mRNA, is also possible, but as it requires similar large amounts of RNA to perform this is typically prohibited by cost.

Based on these descriptions it is clear that electroporation has an advantage over microinjection for the introduction of genome editing nuclease DNA or RNA. But the major disadvantage of the procedure is that the inheritance of the electroporated DNA can be somewhat unpredictable leading to mosaic expressions of the construct. This requires that the investigator thoroughly account for possible mosaicism when reaching conclusions on data. The inheritance of microinjected media is more reliably inherited equally between cells, and as previously mentioned can be injected into specific blastomeres, but for lineage specific genome editing electroporation is a more reliable method. If an investigator wishes to disrupt a gene expressed in the tail muscle cells only, and not in other mesodermal tissues, for example the notochord, the injection media would have to be introduced into the B7.4, B7.8 b7.17, A8.16 and B8.10 cells in both the left and right hemispheres of the embryo. Even if the investigator could do this for one embryo, by the time the work was completed the other embryos would no longer be in the 64-cell stage prohibiting another injection to be attempted. The electroporation of a site-specific nuclease whose expression was driven by the promoter of a muscle marker gene such as *Ci-TnI* would allow the nuclease to be expressed in the cell lineages described above, and in their descendants throughout the larval stages of development. However there are disadvantages to electroporation in addition to the above-mentioned mosaicism. If one wishes to study very early development, up to the 8-cell stage, *Ciona* embryos are zygoticaly silent; therefore the introduction of DNA would be useless. However in this case the introduction of a site-specific nuclease may also be unreliable as it is unlikely to act quickly enough to achieve the required result. In this case either a multigenerational gene knockout using genome-editing strategy would work. But it is probably better to use MASOs at this stage. From the 16-cell stage to the late gastrula stage mRNA injection would probably be preferable to electroporation as early expression would be important. If the knockout is required at the larval stage or during and throughout metamorphosis then a conditional knockout using an electroporated construct would probably be suitable.

7.3 Genome Editing in *Ciona*

7.3.1 ZFNs in Ciona

There has been a single report describing the use of zinc finger nucleases (ZFNs) in *Ciona* (Kawai et at 2012). Using a transgenic line with a stably incorporated eGFP gene expressed in muscle cell precursors and larval tail muscle, mRNA for a ZFN pair was injected into *Ciona* eggs and allowed to develop. At the larval stage injected individuals could be seen with reduced eGFP expression levels. As the eGFP gene is present in the genome and is either expressed, or not expressed depending on whether or not the sperm that fertilized the egg contained a chromosome with the transgene, this disruption must be due to the action of the ZFNs. Differing amounts of ZFN mRNA was injected resulting in an increase in mutational efficiency reaching 100 % when 150 fg was injected. Notable toxicity only appeared when 20 times this amount of mRNA was injected. This is the highest rate of mutational efficiency relative to toxicity in either our published or unpublished reports. Therefore it may be sensible to conclude that based on this very limited dataset of a single transgene, ZFNs are the best choice for genome editing in ascidians. However we currently have no ongoing work using ZFNs and no plans to restart this work. Based on our observations the use of ZFNs is rapidly declining due to the difficulty in constructing a suitably active pair of ZFNs for genome editing. Comparing the other two methods of nuclease construction to ZFNs, TALENs are comparatively easy to make and CRISPR/Cas9 is trivial.

7.3.2 TALENs in Ciona

We described the virtues of electroporation-mediated transfection of *Ciona* embryos already in this chapter. A standard procedure in TALEN construction is the ligation of section of DNA corresponding to a single TAL domain that binds to a single DNA nucleotide. If the assembly is done into a plasmid that has been pre-designed with a promoter and regulatory elements that can express the gene in *Ciona* all that is needed is to prepare a large enough quantity of DNA for electroporation. However there is a weakness to this approach that the electroporation can be somewhat unreliable and can also result in a variable proportion of poorly developing embryos due to the many treatments that they are exposed to and the potential toxicity of the introduced constructs. To overcome this we constructed TALENs that contained a reporter cassette that expressed bright fluorescent proteins (Treen et al. 2014). This was initially designed to express in the epidermis and the TALENs were expressed using an *EF1α* promoter to express TALENs in all transcriptionally active cells. When the genomic DNA of these electroporations was isolated from embryos showing high consistent expression of the fluorescent proteins, mutation rates between 71 % and 95 % were observed. This indicates that TALENs can be a robust tool for performing knockouts in G0 embryos extremely quickly. Substituting the ubiquitous *EF1α* promoter for one of the many promoters can drive transgene expression in defined cell lineages, often tissue specific, for example the *Brachyury* promoter is exclusively expressed in notochord precursor cells by the 112 cell stage (Corbo et al. 1997), and the *Snail* promoter is exclusively expressed in tail muscle precursor cells at the neurula stage (Fujiwara et al. 1998). This precise expression could be used to investigate the function of genes in more detail. This was shown with promoters knocking out *Hox12* in the epidermis and *Fgf3* in early neural cells (Treen et al. 2014). When this was performed, previously described morpholino knockdown phenotypes (Ikuta et al. 2010; Shi et al. 2009) of these genes were successfully observed. Taking this technique further we used a *PC2* promoter that is drives expression in early mature neuronal cells to knock out *fgf3* in the larval stages. When *Fgf3* targeting TALENs were expressed using the *PC2* promoter it caused a previously undescribed defect in metamorphosis where tail reabsorption was halted at approximately halfway through this event. This knockout phenotype could not have been observed with MASOs, TALEN mRNA microinjections or ubiquitously expressed electroporated TALENs due to an earlier function of *Fgf3* in notochord formation. When a similar experiment was performed to knock down *Hox12* in the epidermis, a previously described rounded tail was observed, presumably in part due to the loss of expression of downstream *Fgf8* (Fig. 7.2c,d).

In order to attempt germ-line knockouts using TALENs in *Ciona*, an advancement made possible by advances in ascidian culturing techniques allowing lines to be maintained for successive generations (Joly et al. 2007), microinjections of TALEN mRNA was performed. Although germ-line knockouts may be possible using electroporations, the cell lineage from which the germ-cells originate is transcriptionally inert during embryogenesis (Shirae-Kurabayashi et al. 2011), so the amount of mutants in the G1 generations may be low. We were able to detect TALEN induced mutations in genomic DNA extracted from sperm of TALEN microinjected animals that could be inherited in the G1 animals (Yoshida et al. 2014). When *Hox4* and *Hox5* genes were targeted, no defects in larval or juvenile development could be seen, this is also consistent with previous data from MASO knockdowns (Ikuta et al. 2010).

7.3.3 *CRISPR/Cas9 in* Ciona

To date there have been no successful reports of the use of CRISPR/Cas9 in *Ciona* or other ascidians. This is somewhat surprising considering its supposed ease of use and activity in a wide range of organisms. Very recently we have succeeded in using CRISPR/Cas9 in *Ciona* (Sasaki et al. 2014). This was achieved using both microinjections of short guide RNA (sgRNA) and Cas9 mRNA, as well as electroporations using a U6 promoter to express sgRNA and an $EF1\alpha$ promoter to express the Cas9 nuclease. Targeting *Hox3* and *Hox5* we found that CRISPR/Cas9 can be used in *Ciona*, but its activity appears to be lower than TALENs. However it was capable of targeting loci that TALENs failed to. Toxicity and off-target effects in *Ciona* were very low. Therefore CRISPR/Cas9 is a welcome addition, particularly for recalcitrant genomic loci, but it will require some optimization the advantage of ease of construction is not outweighed by lower mutational activity.

7.3.4 *Potential for Genome Editing in Future Ascidian Research*

The major new approach that genome editing has opened is the possibility to investigate later stages of ascidian development. An overall goal of ascidian research is the ability to unravel ancestral features of chordate development (Passamaneck and Di Gregorio 2005). The possibility that ascidian larvae can be used for such a purpose has been widely accepted, but the clear morphological differences between ascidians and vertebrates, as well as differences in gene expression patterns are difficult to ignore. Genome editing will provide the ability to investigate gene function during and after metamorphosis. The ascidian early tailbud stage has been proposed as the equivalent of the vertebrate phylotypic stage, however we find this difficult to accept as none of the adult, or even larval organs have formed at this stage. Investigations using genome editing can assist in understanding the developmental programs that form ascidian larval, juvenile and adult organs. A comparison between this and what is observed in vertebrates, amphioxus and hemichordates could be used to reverse engineer the ancestral chordate.

Another use that investigations in ascidians can uncover is the elucidation of gene regulatory networks. This could be used to infer the networks of the ancestral vertebrate, but it can also be used to better understand the logical transactions that take place between the genome and the expressed genes that lead to patterning and morphogenesis (Peter and Davidson 2009). While ascidians have extremely similar larvae with the exception of the number of tail muscle cells (Satoh 1994), the spatial expression, particularly of signaling ligands show clear differences between species (Hudson and Yasuo 2008). Genome editing could be used in a number of ways to investigate development between ascidian species that could answer fundamental questions about how organisms develop and evolve. Although not a genome editing technique, TALE or nuclease-inactivated Cas9 proteins fused to activator and repressor domains (Crocker and Stern 2013; Qi et al. 2013) could be exceptionally powerful tools to answer these questions.

Acknowledgments We would like to thank members in Shimoda Marine Research Center, University of Tsukuba for their help to support our research. This study was supported by Grants-in-Aid for Scientific Research from the Japan Society for the Promotion of Science (JSPS) and Ministry of Education, Culture, Sports, Science and Technology (MEXT) to YS. YS was supported by the Toray Science and Technology Grant. Further support was provided by grants from the National Bioresource Project.

References

Alldredge AL, Madin LP (1982) Pelagic tunicates: unique herbivores in the marine plankton. Biosci 32:655–663

Amat F, Keller PJ (2013) Towards comprehensive cell lineage reconstructions in wm complex organisms using light-sheet microscopy. Dev Growth Differ 55:563–578

Appeltans W, Ahyong ST, Anderson G, Angel MV, Artois T, Bailly N, Bamber R, Barber A, Bartsch I, Berta A, Błażewicz-Paszkowycz M, Bock P, Boxshall G, Boyko CB, Brandão SN, Bray RA, Bruce NL, Cairns SD, Chan T-Y, Cheng L, Collins AG, Cribb T, Curini-Galletti M, Dahdouh-Guebas F, Davie PJF, Dawson MN, De Clerck O, Decock W, De Grave S, de Voogd NJ, Domning DP, Emig CC, Erséus C, Eschmeyer W, Fauchald K, Fautin DG, Feist SW, Fransen CHJM, Furuya H, Garcia-Alvarez O, Gerken S, Gibson D, Gittenberger A, Gofas S, Gómez-Daglio L, Gordon DP, Guiry MD, Hernandez F, Hoeksema BW, Hopcroft RR, Jaume D, Kirk P, Koedam N, Koenemann S, Kolb JB, Kristensen RM, Kroh A, Lambert G, Lazarus DB, Lemaitre R, Longshaw M, Lowry J, Macpherson E, Madin LP, Mah C, Mapstone G,

McLaughlin PA, Mees J, Meland K, Messing CG, Mills CE, Molodtsova TN, Mooi R, Neuhaus B, Ng PKL, Nielsen C, Norenburg J, Opresko DM, Osawa M, Paulay G, Perrin W, Pilger JF, Poore GCB, Pugh P, Read GB, Reimer JD, Rius M, Rocha RM, Saiz-Salinas JI, Scarabino V, Schierwater B, Schmidt-Rhaesa A, Schnabel KE, Schotte M, Schuchert P, Schwabe E, Segers H, Self-Sullivan C, Shenkar N, Siegel V, Sterrer W, Stöhr S, Swalla B, Tasker ML, Thuesen EV, Timm T, Todaro MA, Turon X, Tyler S, Uetz P, van der Land J, Vanhoorne B, van Ofwegen LP, van Soest RWM, Vanaverbeke J, Walker-Smith G, Walter TC, Warren A, Williams GC, Wilson SP, Costello MJ (2012) The magnitude of global marine species diversity. Curr Biol 22:2189–2202

Christiaen L, Wagner E, Shi W, Levine M (2009) The Sea Squirt *Ciona intestinalis*. Cold Spring Harb Protoc 12:pdb.emo138

Corbo JC, Levine M, Zeller RW (1997) Characterization of a notochord-specific enhancer from the Brachyury promoter region of the ascidian, *Ciona intestinalis*. Development 124:589–602

Crocker J, Stern DL (2013) TALE-mediated modulation of transcriptional enhancers in vivo. Nat Methods 10:762–767

Davidson EH, Britten RJ (1971) Note on the control of gene expression during development. J Theor Biol 32:123–130

Dehal P, Satou Y, Campbell RK, Chapman J, Degnan B, De Tomaso A, Davidson B, Di Gregorio A, Gelpke M, Goodstein DM, Harafuji N, Hastings KEM, Ho I, Hotta K, Huang W, Kawashima T, Lemaire P, Martinez D, Meinertzhagen IA, Necula S, Nonaka M, Putnam N, Rash S, Saiga H, Satake M, Terry A, Yamada L, Wang H-G, Awazu S, Azumi K, Boore J, Branno M, Chin-Bow S, DeSantis R, Doyle S, Francino P, Keys DN, Haga S, Hayashi H, Hino K, Imai KS, Inaba K, Kano S, Kobayashi K, Kobayashi M, Lee B-I, Makabe KW, Manohar C, Matassi G, Medina M, Mochizuki Y, Mount S, Morishita T, Miura S, Nakayama A, Nishizaka S, Nomoto H, Ohta F, Oishi K, Rigoutsos I, Sano M, Sasaki A, Sasakura Y, Shoguchi E, Shin-i T, Spagnuolo A, Stainier D, Suzuki MM, Tassy O, Takatori N, Tokuoka M, Yagi K, Yoshizaki F, Wada S, Zhang C, Hyatt PD, Larimer F, Detter C, Doggett N, Glavina T, Hawkins T, Richardson P, Lucas S, Kohara Y, Levine M, Satoh N, Rokhsar DS (2002) The draft genome of *Ciona intestinalis*: insights into chordate and vertebrate origins. Science 298:2157–2167

Delsuc F, Brinkmann H, Chourrout D, Philippe H (2006) Tunicates and not cephalochordates are the closest living relatives of vertebrates. Nature 439:965–968

Fujiwara S, Corbo JC, Levine M (1998) The snail repressor establishes a muscle/notochord boundary in the *Ciona* embryo. Development 125:2511–2520

Hudson C, Yasuo H (2008) Similarity and diversity in mechanisms of muscle fate induction between ascidian species. Biol Cell 100:265–277

Ikuta T, Satoh N, Saiga H (2010) Limited functions of Hox genes in the larval development of the ascidian *Ciona intestinalis*. Development 137:1505–1513

Imai KS, Levine M, Satoh N, Satou Y (2006) Regulatory blueprint for a chordate embryo. Science 312:1183–1187

Joly J-S, Kano S, Matsuoka T et al (2007) Culture of *Ciona intestinalis* in closed systems. Dev Dyn 236:1832–1840

Kawai N, Ochiai H, Sakuma T, Yamada L, Sawada H, Yamamoto T, Sasakura Y (2012) Efficient targeted mutagenesis of the chordate *Ciona intestinalis* genome with zinc-finger nucleases. Dev Growth Differ 54:535–545

Kawamura K, Kitamura S, Sekida S, Tsuda M, Sunanaga T (2012) Molecular anatomy of tunicate senescence: reversible function of mitochondrial and nuclear genes associated with budding cycles. Development 139:4083–4093

Kumano G, Negoro N, Nishida H (2014) Transcription factor Tbx6 plays a central role in fate determination between mesenchyme and muscle in embryos of the ascidian, *Halocynthia roretzi*. Dev Growth Differ 56:310–322

Lemaire P (2011) Evolutionary crossroads in developmental biology: the tunicates. Development 138:2143–2152

Nishida H, Satoh N (1983) Cell lineage analysis in ascidian embryos by intracellular injection of a tracer enzyme. I. Up to the eight-cell stage. Dev Biol 99:382–394

Omotezako T, Nishino A, Onuma TA, Nishida H (2013) RNA interference in the appendicularian *Oikopleura dioica* reveals the function of the Brachyury gene. Dev Genes Evol 223:261–267
Passamaneck YJ, Di Gregorio A (2005) *Ciona intestinalis*: chordate development made simple. Dev Dyn 233:1–19
Peter IS, Davidson EH (2009) Genomic control of patterning. Int J Dev Biol 53:707–716
Qi LS, Larson MH, Gilbert LA, Doudna JA, Weissman JS, Arkin AP, Lim WA (2013) Repurposing CRISPR as an RNA-guided platform for sequence-specific control of gene expression. Cell 152:1173–1183
Robin FB, Dauga D, Tassy O, Sobral D, Daian F, Lemaire P (2011) Time-lapse imaging of live *Phallusia* embryos for creating 3D digital replicas. Cold Spring Harb Protoc 2011:1244–1246
Sasaki S, Yoshida K, Hozumi A, Sasakura Y (2014) CRISPR/Cas9-mediated gene knockout in the ascidian *Ciona intestinalis*. Dev Growth Differ 56:499–510
Sasakura Y, Sierro N, Nakai K, Inaba K, Kusakabe TG (2012) Genome structure, functional genomics, and proteomics in ascidians. In: Genome mapping and genomics in laboratory animals. Springer, Berlin, Heidelberg, pp 87–132
Satoh N (1994) Developmental biology of ascidians. Cambridge University Press, New York
Satoh N (2014) Developmental genomics of ascidians. Wiley Brackwell, New York
Satoh N, Araki I, Satou Y (1996) An intrinsic genetic program for autonomous differentiation of muscle cells in the ascidian embryo. Proc Natl Acad Sci U S A 93:9315–9321
Seo HC, Kube M, Edvardsen RB, Jensen MF, Beck A, Spriet E, Gorsky G, Thompson EM, Lehrach H, Reinhardt R, Chourrout D (2001) Miniature genome in the marine chordate *Oikopleura dioica*. Science 294:2506
Shi W, Peyrot SM, Munro E, Levine M (2009) FGF3 in the floor plate directs notochord convergent extension in the *Ciona* tadpole. Development 136:23–28
Shirae-Kurabayashi M, Matsuda K, Nakamura A (2011) Ci-Pem-1 localizes to the nucleus and represses somatic gene transcription in the germline of *Ciona intestinalis* embryos. Development 138:2871–2881
Small KS, Brudno M, Hill MM, Sidow A (2007) A haplome alignment and reference sequence of the highly polymorphic *Ciona savignyi* genome. Genome Biol 8:R41
Stolfi A, Christiaen L (2012) Genetic and genomic toolbox of the chordate *Ciona intestinalis*. Genetics 192:55–66
Treen N, Yoshida K, Sakuma T, Sasaki H, Kawai N, Yamamoto T, Sasakura Y (2014) Tissue-specific and ubiquitous gene knockouts by TALEN electroporation provide new approaches to investigating gene function in *Ciona*. Development 141:481–487
Veeman MT, Chiba S, Smith WC (2011) *Ciona* genetics. In: Vertebrate embryogenesis. Humana Press, Totowa, pp 401–422
Voskoboynik A, Newman AM, Corey DM, Sahoo D, Pushkarev D, Neff NF, Passarelli B, Koh W, Ishizuka KJ, Palmeri KJ, Dimov IK, Keasar C, Fan HC, Mantalas GL, Sinha R, Penland L, Quake SR, Weissman IL (2013a) Identification of a colonial chordate histocompatibility gene. Science 341:384–387
Voskoboynik A, Neff NF, Sahoo D, Newman AM, Pushkarev D, Koh W, Passarelli B, Fan HC, Mantalas GL, Palmeri KJ, Ishizuka KJ, Gissi C, Griggio F, Ben-Shlomo R, Corey DM, Penland L, White RA, Weissman IL, Quake SR (2013b) The genome sequence of the colonial chordate, *Botryllus schlosseri*. Elife 2:e00569
Wada H, Satoh N (1994) Details of the evolutionary history from invertebrates to vertebrates, as deduced from the sequences of 18S rDNA. Proc Natl Acad Sci USA 91:1801–1804
Whittaker JR (1973) Segregation during ascidian embryogenesis of egg cytoplasmic information for tissue-specific enzyme development. Proc Natl Acad Sci U S A 70:2096–2100
Yoshida K, Treen N, Hozumi A et al (2014) Germ cell mutations of the ascidian *Ciona intestinalis* with TALE nucleases. Genesis 52:431–439

Chapter 8
Genome Editing in Zebrafish and Medaka

Atsuo Kawahara, Taijiro Yabe, Satoshi Ansai, Shinji Takada, and Masato Kinoshita

Abstract Two small ornamental fish, zebrafish and medaka, have become popular model vertebrates suitable for genetic analysis. A number of mutant fish defective in organogenesis and biological responses have been isolated from chemical mutagenesis screens and forward genetic analyses of these mutants have identified novel essential regulators involved in morphological and physiological processes. Recently, remarkable innovations in genome editing technologies, such as zinc finger nucleases (ZFNs) and transcription activator-like effector nucleases (TALENs) as well as the clustered interspaced short palindromic repeats (CRISPR)/CRISPR-associated (Cas) system, have enabled us to introduce double-strand DNA breaks at target genomic loci, subsequently leading to the disruption of targeted genes (knockout) or their replacement with homologous fragments (knockin). Here, we summarize the usefulness and application of targeted genome modifications in zebrafish and medaka and their relevance to the basic and medical sciences.

Keywords CRISPR/Cas9 • Genome editing • Medaka • TALEN • Zebrafish

A. Kawahara (✉)
Laboratory for Developmental Biology, Center for Medical Education and Sciences, Graduate School of Medical Science, University of Yamanashi, Simogatou 1111, Chuo, Yamanashi 409-3898, Japan
e-mail: akawahara@yamanashi.ac.jp

T. Yabe • S. Takada
Okazaki Institute for Integrative Bioscience, National Institute of Natural Sciences, Okazaki, Aichi 444-8787, Japan

Department of Basic Biology, The Graduate University for Advanced Studies (SOKENDAI), Okazaki, Aichi 444-8787, Japan

S. Ansai • M. Kinoshita
Division of Applied Biosciences, Graduate School of Agriculture, Kyoto University, Kitashirakawa-Oiwake, Sakyo, Kyoto, 626-8502, Japan

T. Yamamoto (ed.), *Targeted Genome Editing Using Site-Specific Nucleases*,
DOI 10.1007/978-4-431-55227-7_8

Abbreviations

CRISPR	clustered interspaced short palindromic repeats
TALEN	transcription activator-like effector nuclease
ZFN	zinc finger nuclease

8.1 Introduction

Various types of multicellular organisms, such as the fruit fly (*Drosophila*), nematode (*C. elegans*), fish (zebrafish and medaka), frog (*Xenopus*), chicken and mouse, are widely used as model organisms in the life sciences. Among them, both zebrafish and medaka have proven to be convenient model animals for vertebrate genetic analysis (Haffter et al. 1996; Driever et al. 1996; Furutani-Seiki et al. 2004). The conserved molecular structure and function between fish and mammals have allowed zebrafish and medaka studies to be extrapolated to mammals.

Zebrafish (*Danio rerio*) is a tropical small fish that is approximately 4 cm long and has a 3-month generation time (Fig. 8.1a). Its embryos are transparent and develop rapidly, therefore, most organs start to function within several days after fertilization. The transparency, in combination with the expression of fluorescent proteins driven by a tissue-specific promoter, enables the visualization of organogenesis (Fig. 8.1b) (Lawson and Weinstein 2002). A zebrafish genome project has recently revealed that the zebrafish genome contains approximately 26,000 genes (Howe et al. 2013). Seventy percent of human genes exhibit similarity to their counterparts in zebrafish. Thus, forward and reverse genetic analyses of zebrafish could provide important information on human genetic disorders.

Medaka (*Oryzias latipes*) is a temperate freshwater small fish and shares the advantages of zebrafish as a model system: transparency of embryos, short generation time and easy access to its genomic information on the web (Fig. 8.1c). Similar to zebrafish, many mutants (spontaneous and induced) and transgenic strains are available to researchers (National BioResource Project Medaka: http://www.shigen.nig.ac.jp/medaka) (Fig. 8.1d). More than a century of biological study of medaka has produced a number of inbred strains that contain sufficient polymorphism for positional cloning. In addition, wild populations, which may harbor SNPs in each population, are still available, and recently, a project established a near-sub-isogenic panel of wild lines from a single wild population (Spivakov et al. 2014). These genetic variations will contribute to the analysis regarding the correlation of genes, such as modifiers of disease-related genes.

To date, more than two thousand zebrafish mutants defective in organogenesis and biological responses with similar symptoms to human genetic disorders have been isolated from chemical mutagenesis screens using the mutagen *N*-ethyl-*N*-*nitrosourea* (ENU). For example, in humans, a genetic mutation in the T-box transcriptional factor *TBX5* was identified in patients with Holt-Oram syndrome (Basson et al. 1997; Li et al. 1997) that results in heart and limbs abnormalities.

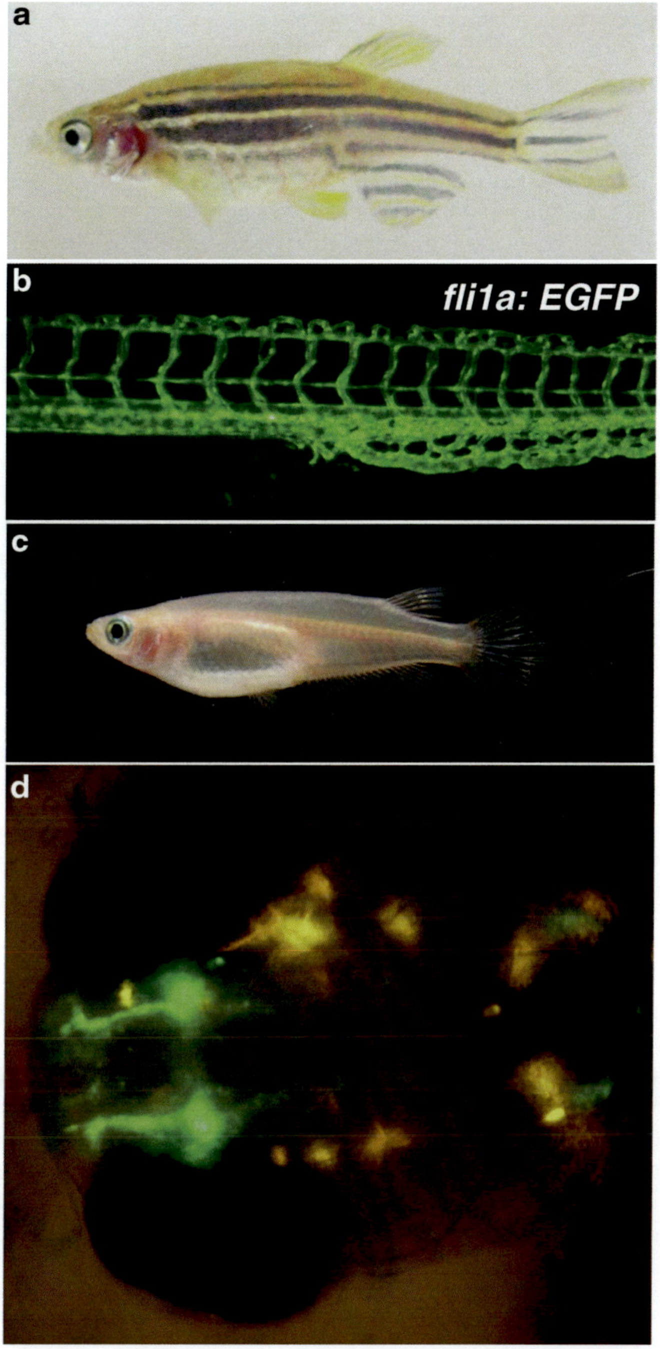

Fig. 8.1 Model vertebrates, zebrafish and medaka. (**a**) Wild-type adult zebrafish. (**b**) Transgenic zebrafish, Tg(*fli1a:EGFP*), expressing enhanced green fluorescent protein (EGFP) driven by the endothelial-specific *fli1a* promoter. (**c**) Wild-type adult medaka: Hd-rR strain. (**d**) Transgenic medaka, Tg(*gnrh3: hrGFP*) expressing hrGFP driven by the *gnrh3* gene promoter. The yellow fluorescence is autofluorescence of the leucophore (provided by Dr. K. Okubo, University of Tokyo). The images show the anterior right

Consistent with this finding, a zebrafish mutant with severe developmental defects in the heart and limbs was shown to be disrupted in the *tbx5* gene (Garrity et al. 2002). Conversely, some genes responsible for human diseases were initially identified in zebrafish or medaka. For example, it was first reported that the gene responsible for the zebrafish anemia mutant *weissherbst* was *ferroportin1*, the gene product of which functions as a basolateral iron exporter (Donovan et al. 2000). Following this study, human *Ferroportin* (*SLC11A3*) was identified as the gene responsible for type IV hemochromatosis, a progressive iron overload disorder (Montosi et al. 2001; Njajou et al. 2001). Similarly, the *ktu* gene, which was initially identified in a medaka induced mutant screening, is conserved among animals and ciliated unicellular organisms (Omran et al. 2008). In this case, the analysis of the medaka *ktu* mutant, which shows loss of ciliary motility leading to situs inversus, led to the identification of mutations in the KTU gene in two consanguineous families of human primary ciliary dyskinesia (PCD) patients, suggesting that KTU mutation is a possible driver of PCD.

In addition to their use as human disease models (Matsui et al. 2010; Morita et al. 2012), the gene knockout method is also a powerful tool in the study of social behavior. The combination of a gene knockout model (TILLING) and electrophysiological and other methods revealed that the TN-GnRH3 neuron, one of three GnRH3 neurons is deeply involved in the mating preferences of female medaka, that is, female medaka prefer familiar males to novel ones (Okuyama et al. 2014). Thus, forward genetic analysis using zebrafish and medaka provides a powerful approach for understanding the molecular mechanisms of various physiological phenomena and human diseases.

8.2 Genome Editing Technologies in Fish

The development of genome editing technologies, such as ZFN, TALEN and CRISPR/Cas9, has made the disruption of targeted genes feasible in any organism, including fish (Hisano et al. 2014) . Both ZFN and TALEN are site-specific artificial nucleases that consist of a DNA binding domain (Zinc finger domain or TALE domain) and a *Fok*I nuclease catalytic domain (Fig. 8.2a). At present, TALEN is more popular because of its predictable target specificity and ease of construction. We usually use four different types of TALE repeats, which contain an identical 34 amino acid sequence, except for two hypervariable residues at amino acid positions 12 and 13. The TALE repeats recognize one specific nucleotide (TALE code: NG type=T, HD type=C, NI type=A, NN type or NH type=G). Because forward- and reverse-TALENs recognize 20–30 nucleotides, they provide a highly specific interaction with the targeted genomic locus and low off-target effects.

CRISPR/Cas9 consists of two separate components, the RNA-guided nuclease Cas9 and guide RNA (gRNA), which contains 20 base complementary sequences against the targeted genomic site (Fig. 8.2b) (Cong et al. 2013; Mali et al. 2013). The design of gRNA is flexible because the gRNA target sequence is followed by a

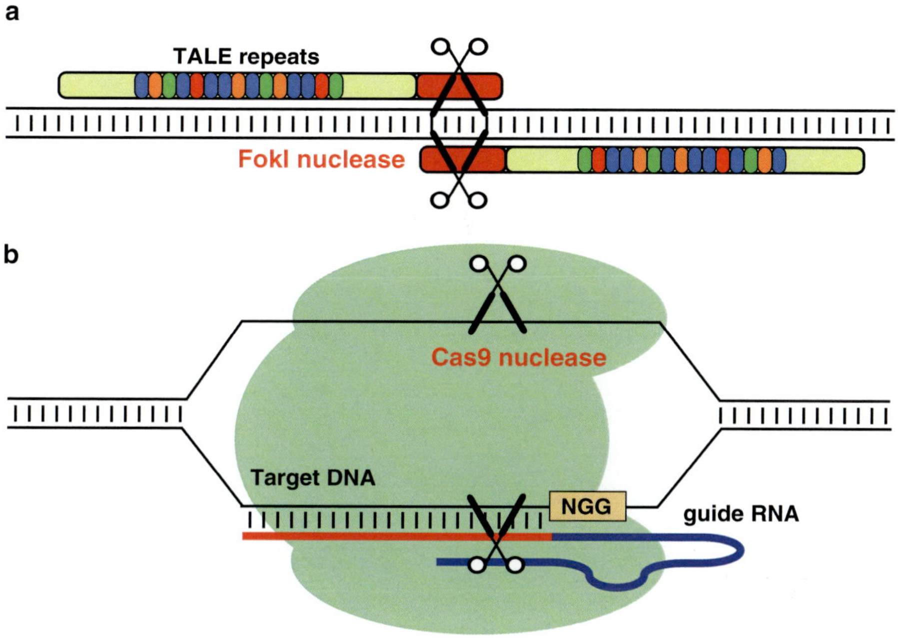

Fig. 8.2 Two genome editing technologies. (**a**) TALEN. Each TALE repeat recognizes one nucleotide. Forward- and reverse-TALEN induce double-strand DNA breaks at the target site, which cause framesfit mutations. (**b**) CRISPR/Cas9. The guide RNA (gRNA) recognizes 20 nucleotides at the target site, and the Cas9 nuclease generates double-strand DNA breaks just inside the PAM sequence (NGG, N = any nucleotide)

protospacer-adjacent motif (PAM) sequence NGG (N: any nucleotide), and the construction of a plasmid that produces gRNA is quite easy compared to TALEN formation. There are some concerns about off-target effects in the CRISPR/Cas9 system because of the very short recognition length (20 bases) of the gRNA. However, side effects could be diminished by out-crossing to wild-type. Additionally, the CRISPR/Cas9 system appeared to be suitable for multiple genome modifications because multiple gRNAs work well with the nuclease Cas9, as described below.

Both TALEN and CRISPR/Cas9 introduce double-strand DNA breaks at the target site. These breaks are repaired by one of two major pathways: non-homologous end joining (NHEJ) and homologous recombination (HR). NHEJ can join the broken DNA strands without any templates, frequently leading to a small insertion and/or deletion (indel mutation). Because indel mutations often cause frameshifts in the open reading frame of the targeted gene, properly targeted knockout fish can be easily identified by screening for the mutation. In the presence of homologous template DNAs, double-strand DNA breaks can be repaired through the replacement of flanking homology arms (knockin) using the HR system.

We have established the efficient generation of knockout zebrafish and medaka (Fig. 8.3). One key to this system is the estimation of the efficacy of the genome

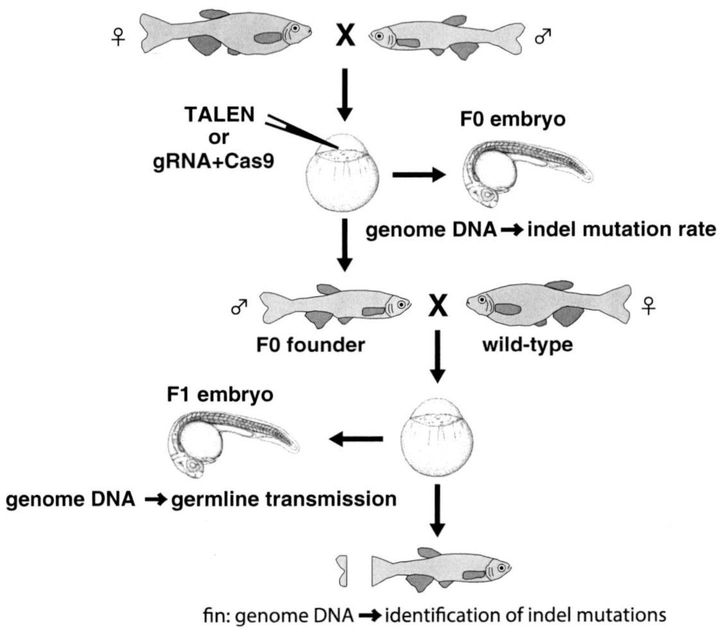

Fig. 8.3 Strategy for establishing targeted gene knockout fish. TALEN or gRNA/Cas9 is injected into zebrafish or medaka embryos. The genome editing activity is estimated by examining the indel mutation rate in the genomic DNA from the TALEN- or gRNA/Cas9-injected embryos. Adult F0 founders are mated with wild-type fish, and genomic DNAs from F1 embryos are prepared. Both the germline transmission in F1 embryos and the identification of indel mutations in growing F1 fish are investigated by the heteroduplex mobility assay (HMA; see Fig. 8.4)

editing activities. We have found that the heteroduplex mobility assay (HMA) is very useful for the identification of targeted genome modifications (Ota et al. 2013). Because TALEN and CRISPR/Cas9 predominantly induce small indel mutations, both heteroduplexes and homoduplexes are produced in the PCR amplification of the targeted genomic site. The principal of HMA is very simple in that heteroduplexes move more slowly than homoduplexes on polyacrylamide gel electrophoresis because of their opened single-strand structure. More recently, we noted that multiple genome modifications induced by gRNA/Cas9 could be simultaneously visualized by multi-locus HMA (Fig. 8.4A) (Ota et al. 2014). We have designed

Fig. 8.4 (continued) MCE-202 MultiNA (Shimadzu, Kyoto). Amplicons of F2 embryos obtained by intercrossing heterozygous mutant F1 fish harboring an 8 bp deletion in the DJ-1 gene were subjected to HMA with MultiNA. Gel images of HMA of the standard procedure (indicated as "normal HMA"). Heterozygous mutants showed a multiple band pattern (*a*, *c*, *f*, and *h*). Wild-type individuals (*d* and *g*) showed a single band, which moved slightly slower than that of the homozugous mutants (*b*, *e*, and *i*). (**D**) Subsequent HMA to distinguish wild-type and homozygous mutants. An amplicon, which was separately prepared with wild-type DNA, was added to the amplicon of samples that showed a single band in the first HMA (*b*, *d*, *e*, *g*, and *i*). Then, they were reannealed and analyzed with MultiNA. The wild-type samples still showed a single band (*d*' and *g*'), whereas the homozygous mutants (*b*', *e*', and *i*') showed multiple bands

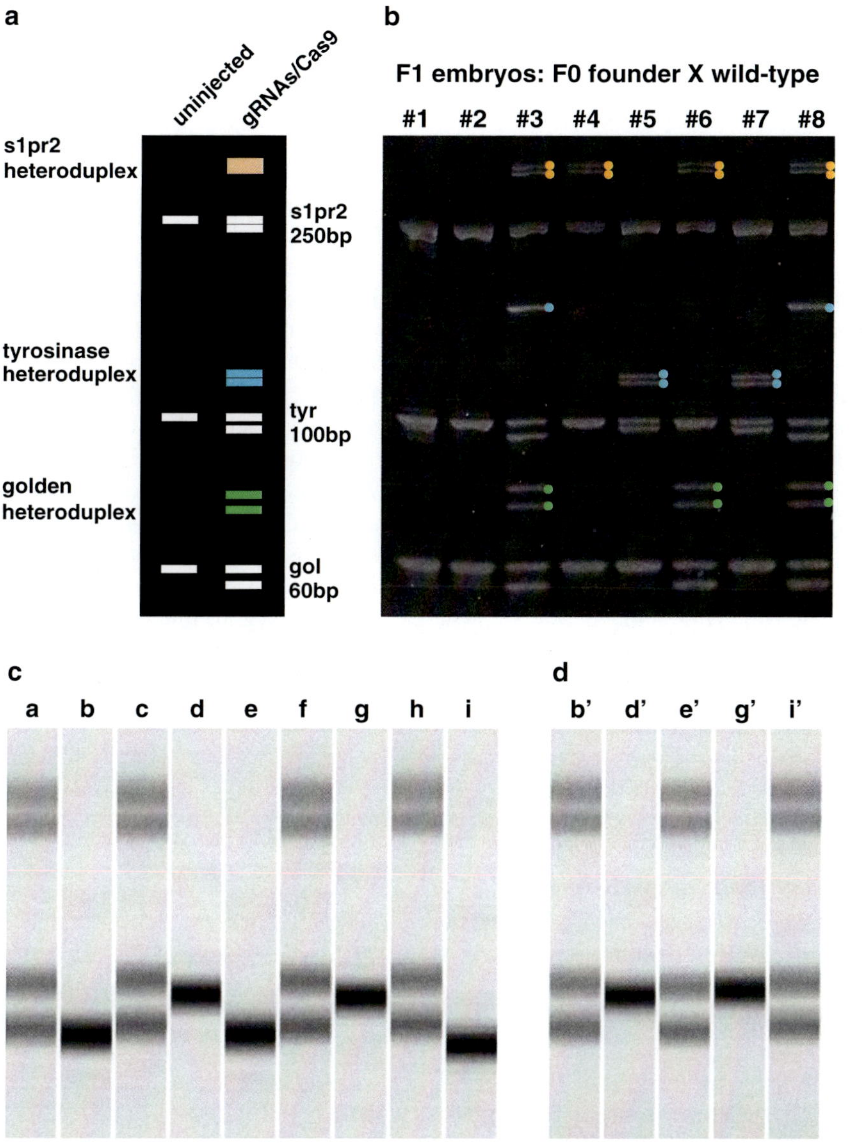

Fig. 8.4 Detection of indel mutations induced by genome editing technologies. (**A**) schematic representation of multi-locus HMA. Individual target sites are amplified using locus-specific primers. A mixture of PCR amplicons (s1pr2, 250 bp; tyrosinase, 100 bp; and golden 60 bp) are separated on a 15 % polyacrylamide gel. Both wild-type and deletion type homoduplexes are indicated by individual white bands. The s1pr2-, tyrosinase- and golden-heteroduplexes are indicated by *orange*, *blue* and *green* bands, respectively. (**B**) Multi-locus HMA in F1 embryos. The F0 founder was mated to wild-type, and multi-locus HMA was performed. Embryos #1 and #2 exhibited no heteroduplexes for any of the target loci. Embryo #4 contains the s1pr2-heteroduplex; embryos #5 and #7 have the tyrosinase-heteroduplex; embryo #6 has the s1pr2- and golden-heteroduplexes; and embryos #3 and #8 have the s1pr2-, tyrosinase- and golden-heteroduplexes, as reported previously (Ota et al. 2014) . (**C**) A gel image of HMA with an automatic electrophoresis system,

locus specific primers positioned at distinct intervals of the individual target sites (e.g., 60 bp, 100 bp and 250 bp) and we succeeded in identifying locus-specific heteroduplexes against three distinct genomic loci on the same gel. Figure 8.4B shows two of eight F1 embryos, #3 and #8, that contain three different indel mutations at these loci; the others contain one or two heteroduplexes (#4, s1pr2-heteroduplex; #5, tyr-heteroduplex; #7, tyr-heteroduplex; #6, s1pr2- and golden-heteroduplex). These results suggest that multi-locus HMA is very useful for detecting multiple genome modifications. If many amplicons are to be subjected to HMA at a time or if routine mutant screening is required, an automatic electrophoresis system is convenient (Ansai et al. 2014) . F2 embryos derived from a F1 heterozygous mutant pair were subjected to HMA with an automatic electrophoresis system, MCE-202 MultiNA (Shimadzu, Kyoto). F2 heterozygous mutants showed multiple bands (Fig. 8.4C: a, c, f, and h). In contrast, wild-type (Fig. 8.4C: d and g) and homozygous mutant (Fig. 8.4C: b, e, and i) individuals showed a single band with different mobilities. As an option, to clarify the difference between wild-type and homozygous mutant embryos, subsequent HMA analysis is effective as follow: Separately prepared amplicon of the wild-type individual is added to the amplicons of both possible wild-type and heterozygous mutant embryos. After reannealing, HMA is performed with MCE-202 MultiNA again. In the second HMA, the wild-type shows a single band (Fig. 8.4D: d' and g'), whereas the homozygous mutant shows multiple bands (Fig. 8.4D: b', e', and i'). As above, the automatic electrophoresis system facilitates mutant screening by saving laborious work and time.

8.3 Multiple Genome Modifications in Zebrafish

Because gRNA is a very small molecule and is separated from the RNA-guided nuclease Cas9, it has been proposed that the CRISPR/Cas9 system is suitable for multiple genome modifications (Jao et al. 2013). We designed five gRNAs for four target genes (*tyrosinase/tyr*, *golden/gol*, *spns2* and *sphingosine-1-phosphate receptor-2/s1pr2*) (Ota et al. 2014). Both *tyr* and *gol* are involved in pigment formation (Page-McCaw et al. 2004; Lamason et al. 2005), while *spns2* and *s1pr2* regulate cardiac progenitor migration (Kupperman et al. 2000; Osborne et al. 2008; Kawahara et al. 2009). When we injected the 5 gRNAs with Cas9 mRNA into zebrafish embryos (F0 embryos), we often observed two distinct phenotypes, the formation of two-separated hearts and pigmentation defects in the skin and retina. We confirmed that the 5 gRNAs/Cas9 injection into zebrafish embryos induced the disruption of multiple targeted genes in somatic cells. We later in-crossed potential F0 founders to acquire F1 embryos. F1 embryo #4 exhibited two morphological defects, the two hearts phenotype and hypopigmentation in the retina, while the F1 embryos #2 and #3 showed the two hearts phenotype but had normal pigmentation in the retina (Fig. 8.5a–h). Sequencing analysis confirmed that individual F1 embryos contained

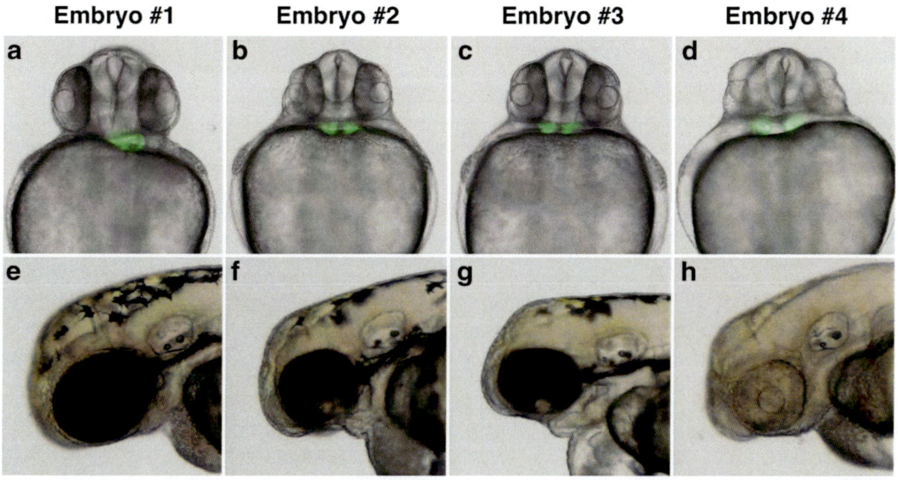

i

gRNA-locus	s1pr2-gRNA	spns2-gRNA1	spns2-gRNA2	gol-gRNA	tyr-gRNA
Embryo #1	WT/WT	WT/WT	WT/+3	WT/WT	WT/WT
Embryo #2	WT/WT	WT/-2	WT/-4(-9/+5)	WT/-4	WT/-6
Embryo #3	WT/-2(-4/+2)	+4(-4/+8)/+10(-6/+16)	WT/WT	WT/-4	WT/-3
Embryo #4	WT/-2(-4/+2)	-2/+10(-6/+16)	WT/-9	-4/-4	-6/-9

Fig. 8.5 Genome editing in zebrafish. Both *golden* and *tyrosinase* are involved in pigment formation in zebrafish, while *s1pr2* and *spns2* regulate cardiac progenitor migration. Five gRNAs (s1pr2-gRNA, spns2-gRNA1, spns2-gRNA2, tyrosinase-gRNA and golden-gRNA) were injected into zebrafish embryos and the embryos were grown into adult F0 founders. F1 embryos were obtained from the in-crossing of F0 founders, and their morphogenic phenotypes were examined (Ota et al. 2014). (**a–d**) Cardiac phenotype. (**e–h**) Pigmentation in the retina. F1 embryo #1 had normal development (**a**, **e**). F1 embryos #2 (**b**, **f**) and #3 (**c**, **g**) exhibited the two hearts phenotype and normal pigmentation in the retina. F1 embryo #4 exhibited the two hearts phenotype and pigmentation defects in the retina (**d**, **h**). (**i**) Genotyping of embryos #1-#4 for individual gRNA-target genomic loci

distinct types of disruption to multiple genes (Fig. 8.5i). These results indicate that CRISPR/Cas9 is a powerful tool for multiple genome modifications in fish.

8.4 Estimation of Deletion Pattern

Microhomology-mediated end joining (MMEJ), a double-strand break (DSB) repair pathway, uses micro-homologous sequences to align the broken ends before joining, thereby causing deletions flanking the original break (McVey and Lee 2008). In the course of our TALEN-mediated gene-targeting experiments with medaka, we noticed

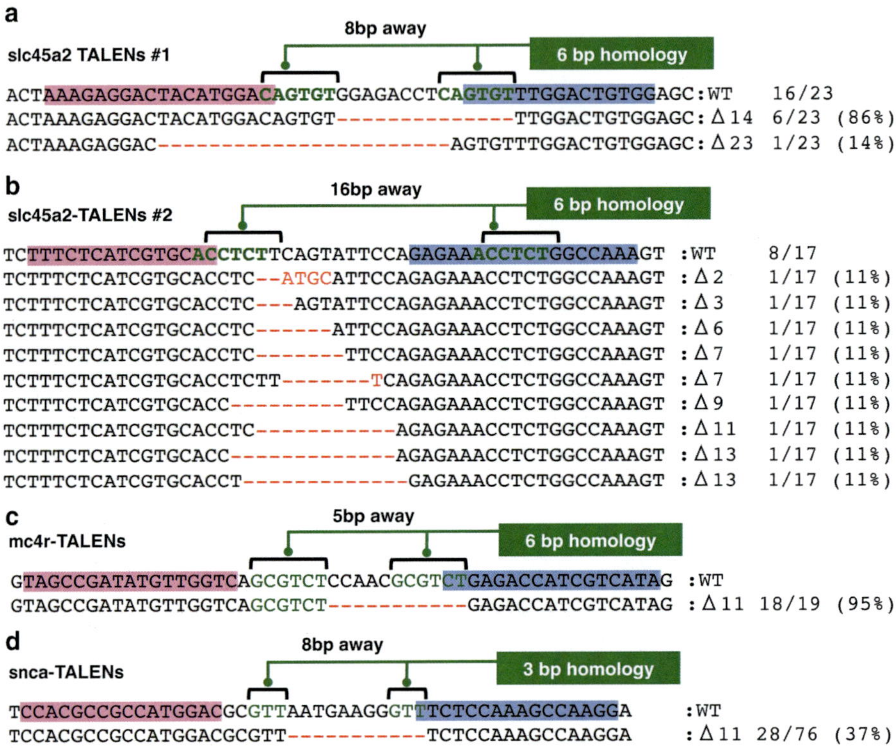

Fig. 8.6 Genome editing in medaka. Mutation pattern analysis in medaka. Subcloned sequences observed in a mixture of genomic DNA extracted from the 12 embryos injected with slc45a2-TALENs for site #1 (**a**) and #2 (**b**). (**a**) Six-base homologous sequences were observed with an 8-base spacer. Six of seven mutants (86 %) had 14-base deletions containing a homologous sequence and spacer, suggesting that MMEJ repair preferentially occurred. (**b**) Six-base homologous sequences were observed with a 16-base spacer. No preferential mutation pattern was observed. Germline mutation sequences induced by snca-TALENs (**c**) and mc4r-TALENs (**d**). The MMEJ repair pathway seems to be preferentially activated with longer homologous sequences (**c**: 6 base) as opposed to shorter homologous sequences (**d**: 3 base). The *red dashes* and *letters* indicate the identified mutations. The *left and right TALEN-binding sequences* are indicated in the *pink* and *blue boxes*, respectively. The *green letters* indicate homologous sequences in the targeting site

that MMEJ is predominately employed in the DSB repair process with specially designed TALEN sequences.

The TALENs, which contain micro-homology sequences (more than 3-nt) spanning over the DSB, preferentially delete the micro-homologous sequence and a spacer sequence between the micro-homologous sequences (Fig. 8.6a, c and d). Relatively long spacer sequences (16-ht) between the micro-homologous sequences, however, did not preferably induce MMEJ (Fig. 8.6b). These findings enable us to predict a mutation pattern (deletion pattern) and facilitate effective gene destruction (Ansai et al. 2014).

8.5 Reduction of Off-Target Alteration

Off-target alteration is a concern with the CRISPR/Cas9 system. For 3 gRNAs (No. 1 and 2a of DJ-1), we screened candidate off-target sequences with a medaka genome database (medaka off-target prediction site: "Pattern Match-CRISPR" http://viewer.shigen.info/medakavw/crisprtool) and found 4 and 8 candidate sites, respectively. Sequence analysis revealed that two of the 8 DJ-1-2a candidates were mutated but that the off-target mutation efficiency was reduced by lowering the concentration of injected gRNAs (Ansai and Kinoshita 2014) .

8.6 Other Applications of TALEN and CRISPR/Cas9

We focus on the techniques for simple mutagenesis mediated by small indel mutations introduced by TALENs and the CRISPR/Cas9 system. Recently, several new techniques have been reported to edit the zebrafish genome more conveniently. Targeted insertion of short DNA sequences containing the loxP site and HA-epitope tag at a targeted genomic locus by the co-injection of single-strand oligonucleotides has been reported (Bedell et al. 2012). The targeted integration of homologous fragments by HR in zebrafish has also been established (Zu et al. 2013), but the efficacy is quite low. The site specific integration of exogenous DNA mediated by homology independent DNA repair using CRISPR/Cas9 and TALEN was also reported (Auer et al. 2014). Paired CRISPR/Cas9- or TALEN-mediated introduction of large deletions and genomic inversions might be powerful tools to edit large regions of small teleost genomes (Ota et al. 2014; Jao et al. 2013; Xiao et al. 2013).

Conclusion

In this chapter, we highlighted how zebrafish and medaka are elegant model vertebrates suitable for forward and reverse genetics. Recent innovations in genome editing technologies have accelerated the elucidation of the physiological function of uncharacterized genes by reverse genetics. Knockin methods using homologous templates in zebrafish and medaka have been hampered by low efficacy. The improvement of knockin technology may enable us to generate human disease model fish by exchanging the mutated genes associated with human genetic disorders with the respective fish counterparts. Because both zebrafish and medaka embryos are very small in size (approximately 1–2 mm), chemical screening using small compounds is suitable for the identification of drug candidates to suppress phenotypic defects in fish models. Therefore, we believe that zebrafish and medaka provide not only understanding of fundamental life science concepts but also valuable insight for the medical sciences.

Acknowledgements We would like to thank Drs. Ota S. and Hisano Y. for valuable discussion. This work was supported by the Funding Program for Next Generation World-Leading Researchers (NEXT Program) and by the Japan Society for the Promotion of Science (JSPS), and a Grant-in-Aid for JSPS Fellows (S.A.).

References

Ansai S, Kinoshita M (2014) Targeted mutagenesis using CRISPR/Cas system in medaka. Biol Open 3:362–371

Ansai S, Inohaya K, Yoshiura Y, Schartl M, Uemura N, Takahashi R, Kinoshita M (2014) Design, evaluation, and screening methods for efficient targeted mutagenesis with transcription activator-like effector nucleases in medaka. Dev Growth Differ 56:98–107

Auer TO, Duroure K, De Cian A, Concordet JP, Del Bene F (2014) Highly efficient CRISPR/Cas9-mediated knock-in in zebrafish by homology-independent DNA repair. Genome Res 24:142–153

Basson CT, Bachinsky DR, Lin RC, Levi T, Elkins JA, Soults J, Grayzel D, Kroumpouzou E, Traill TA, Leblanc-Straceski J et al (1997) Mutations in human TBX5 [corrected] cause limb and cardiac malformation in Holt-Oram syndrome. Nat Genet 15:30–35

Bedell VM, Wang Y, Campbell JM, Poshusta TL, Starker CG, Krug RG 2nd, Tan W, Penheiter SG, Ma AC, Leung AY et al (2012) In vivo genome editing using a high-efficiency TALEN system. Nature 491:114–118

Cong L, Ran FA, Cox D, Lin S, Barretto R, Habib N, Hsu PD, Wu X, Jiang W, Marraffini LA et al (2013) Multiplex genome engineering using CRISPR/Cas systems. Science 339:819–823

Donovan A, Brownlie A, Zhou Y, Shepard J, Pratt SJ, Moynihan J, Paw BH, Drejer A, Barut B, Zapata A et al (2000) Positional cloning of zebrafish ferroportin1 identifies a conserved vertebrate iron exporter. Nature 403:776–781

Driever W, Solnica-Krezel L, Schier AF, Neuhauss SC, Malicki J, Stemple DL, Stainier DY, Zwartkruis F, Abdelilah S, Rangini Z et al (1996) A genetic screen for mutations affecting embryogenesis in zebrafish. Development 123:37–46

Furutani-Seiki M, Sasado T, Morinaga C, Suwa H, Niwa K, Yoda H, Deguchi T, Hirose Y, Yasuoka A, Henrich T et al (2004) A systematic genome-wide screen for mutations affecting organogenesis in medaka, *Oryzias latipes*. Mech Dev 121:647–658

Garrity DM, Childs S, Fishman MC (2002) The heartstrings mutation in zebrafish causes heart/fin Tbx5 deficiency syndrome. Development 129:4635–4645

Haffter P, Granato M, Brand M, Mullins MC, Hammerschmidt M, Kane DA, Odenthal J, van Eeden FJ, Jiang YJ, Heisenberg CP et al (1996) The identification of genes with unique and essential functions in the development of the zebrafish, *Danio rerio*. Development 123:1–36

Hisano Y, Ota S, Kawahara A (2014) Genome editing using artificial site-specific nucleases in zebrafish. Dev Growth Differ 56:26–33

Howe K, Clark MD, Torroja CF, Torrance J, Berthelot C, Muffato M, Collins JE, Humphray S, McLaren K, Matthews L et al (2013) The zebrafish reference genome sequence and its relationship to the human genome. Nature 496:498–503

Jao LE, Wente SR, Chen W (2013) Efficient multiplex biallelic zebrafish genome editing using a CRISPR nuclease system. Proc Natl Acad Sci U S A 110:13904–13909

Kawahara A, Nishi T, Hisano Y, Fukui H, Yamaguchi A, Mochizuki N (2009) The sphingolipid transporter spns2 functions in migration of zebrafish myocardial precursors. Science 323: 524–527

Kupperman E, An S, Osborne N, Waldron S, Stainier DY (2000) A sphingosine-1-phosphate receptor regulates cell migration during vertebrate heart development. Nature 406:192–195

Lamason RL, Mohideen MA, Mest JR, Wong AC, Norton HL, Aros MC, Jurynec MJ, Mao X, Humphreville VR, Humbert JE et al (2005) SLC24A5, a putative cation exchanger, affects pigmentation in zebrafish and humans. Science 310:1782–1786

Lawson ND, Weinstein BM (2002) In vivo imaging of embryonic vascular development using transgenic zebrafish. Dev Biol 248:307–318

Li QY, Newbury-Ecob RA, Terrett JA, Wilson DI, Curtis AR, Yi CH, Gebuhr T, Bullen PJ, Robson SC, Strachan T et al (1997) Holt-Oram syndrome is caused by mutations in TBX5, a member of the Brachyury (T) gene family. Nat Genet 15:21–29

Mali P, Yang L, Esvelt KM, Aach J, Guell M, DiCarlo JE, Norville JE, Church GM (2013) RNA-guided human genome engineering via Cas9. Science 339:823–826

Matsui H, Taniguchi Y, Inoue H, Kobayashi Y, Sakaki Y, Toyoda A, Uemura K, Kobayashi D, Takeda S, Takahashi R (2010) Loss of PINK1 in medaka fish (*Oryzias latipes*) causes late-onset decrease in spontaneous movement. Neurosci Res 66:151–161

McVey M, Lee SE (2008) MMEJ repair of double-strand breaks (director's cut): deleted sequences and alternative endings. Trends Genet 24:529–538

Montosi G, Donovan A, Totaro A, Garuti C, Pignatti E, Cassanelli S, Trenor CC, Gasparini P, Andrews NC, Pietrangelo A (2001) Autosomal-dominant hemochromatosis is associated with a mutation in the ferroportin (SLC11A3) gene. J Clin Invest 108:619–623

Morita A, Nakahira K, Hasegawa T, Uchida K, Taniguchi Y, Takeda S, Toyoda A, Sakaki Y, Shimada A, Takeda H et al (2012) Establishment and characterization of Roberts syndrome and SC phocomelia model medaka (*Oryzias latipes*). Dev Growth Differ 54:588–604

Njajou OT, Vaessen N, Joosse M, Berghuis B, van Dongen JW, Breuning MH, Snijders PJ, Rutten WP, Sandkuijl LA, Oostra BA et al (2001) A mutation in SLC11A3 is associated with autosomal dominant hemochromatosis. Nat Genet 28:213–214

Okuyama T, Yokoi S, Abe H, Isoe Y, Suehiro Y, Imada H, Tanaka M, Kawasaki T, Yuba S, Taniguchi Y et al (2014) A neural mechanism underlying mating preferences for familiar individuals in medaka fish. Science 343:91–94

Omran H, Kobayashi D, Olbrich H, Tsukahara T, Loges NT, Hagiwara H, Zhang Q, Leblond G, O'Toole E, Hara C et al (2008) Ktu/PF13 is required for cytoplasmic pre-assembly of axonemal dyneins. Nature 456:611–616

Osborne N, Brand-Arzamendi K, Ober EA, Jin SW, Verkade H, Holtzman NG, Yelon D, Stainier DY (2008) The spinster homolog, two of hearts, is required for sphingosine 1-phosphate signaling in zebrafish. Curr Biol 18:1882–1888

Ota S, Hisano Y, Muraki M, Hoshijima K, Dahlem TJ, Grunwald DJ, Okada Y, Kawahara A (2013) Efficient identification of TALEN-mediated genome modifications using heteroduplex mobility assays. Genes Cells 18:450–458

Ota S, Hisano Y, Ikawa Y, Kawahara A (2014) Multiple genome modifications by the CRISPR/Cas9 system in zebrafish. Genes Cells 19:555–564

Page-McCaw PS, Chung SC, Muto A, Roeser T, Staub W, Finger-Baier KC, Korenbrot JI, Baier H (2004) Retinal network adaptation to bright light requires tyrosinase. Nat Neurosci 7:1329–1336

Spivakov M, Auer TO, Peravali R, Dunham I, Dolle D, Fujiyama A, Toyoda A, Aizu T, Minakuchi Y, Loosli F, et al. (2014) Genomic and phenotypic characterization of a wild medaka population: towards the establishment of an isogenic population genetic resource in fish. G3 (Bethesda) 4:433–445

Xiao A, Wang Z, Hu Y, Wu Y, Luo Z, Yang Z, Zu Y, Li W, Huang P, Tong X et al (2013) Chromosomal deletions and inversions mediated by TALENs and CRISPR/Cas in zebrafish. Nucleic Acids Res 41:e141

Zu Y, Tong X, Wang Z, Liu D, Pan R, Li Z, Hu Y, Luo Z, Huang P, Wu Q et al (2013) TALEN-mediated precise genome modification by homologous recombination in zebrafish. Nat Methods 10:329–331

Chapter 9
Genome Editing Using Site-Specific Nucleases in Amphibians

Ken-ichi T. Suzuki and Toshinori Hayashi

Abstract Amphibians have a long history as model animals and have greatly contributed to biological research fields, especially developmental biology and cell biology, including embryonic induction, signal transduction, pattern formation, cell cycle regulation, nuclear reprogramming, metamorphosis, and organ regeneration. In addition to the historical achievements, recent advances in genome editing using site-specific nucleases have facilitated reverse genetics research targeting genes of interest in amphibians. The epochal tool enables the performance of not only knockout of genes of interest, but also knockin of genes into particular target genomic loci, which was never previously possible in amphibians. Here, we review recent studies involving genome editing with zinc finger nucleases (ZFNs), transcription activator-like effector nucleases (TALENs), and clustered regularly interspaced short palindromic repeats (CRISPR)/CRISPR-associated (Cas) in the amphibians *Xenopus laevis (frog)*, *Xenopus (Silurana) tropicalis (frog)*, *Pleurodeles waltl (newt)*, and *Ambystoma mexicanum (axolotl)*, all of which are known to be excellent model animals in developmental biology and regeneration biology. We also discuss their possibilities as model animals when carrying such a robust reverse genetics tool.

Keywords Ambystoma mexicanum • CRISPR/Cas9 • Knockin • Knockout • Multiplex gene disruption • Pleurodeles waltl • TALENs • Xenopus (Silurana) tropicalis • Xenopus laevis • ZFNs

K.-i.T. Suzuki (✉)
Department of Mathematical and Life Sciences, Graduate School of Science, Hiroshima University, 1-3-1 Kagamiyama, Higashi-Hiroshima, Hiroshima 739-8526, Japan
e-mail: suzuk107@hiroshima-u.ac.jp; suzuken107@gmail.com

T. Hayashi (✉)
Department of Biomedical Sciences, Faculty of Medicine, School of Life Science, Tottori University, 86 Nishi-cho, Yonago, Tottori 683-8503, Japan
e-mail: toshi2@med.tottori-u.ac.jp

T. Yamamoto (ed.), *Targeted Genome Editing Using Site-Specific Nucleases*,
DOI 10.1007/978-4-431-55227-7_9

9.1 Introduction

To date, many amphibians have provided excellent contributions to basic biology, especially developmental biology and cell biology, because they have some exceptional advantages as experimental animals compared with other species. First, their eggs and embryos are large and can be abundantly obtained relative to other animals. Second, their developmental processes can be clearly observed through the whole lifespan. Third, their embryos can successfully endure surgical manipulation and transplantation. In this chapter, we introduce genome editing research in two anurans and three urodeles that are widely used as model animals in developmental biology and regeneration biology (Fig. 9.1).

Among the amphibians, *Xenopus laevis* is a useful model animal for studying developmental biology and cell biology, including embryonic induction, signal transduction, pattern formation, cell cycle regulation, nuclear reprogramming, metamorphosis, and regeneration, because researchers can easily and abundantly obtain eggs all year round by injecting chorionic gonadotropin (Harland and Grainger 2011). Robust eggs and embryos enable the injection of mRNAs and antisense oligonucleotides targeting genes of interest to evaluate their in vivo functions. Large amounts of egg extracts can be applied to cell biology for cell cycle regulation using biochemical approaches. In addition, transcriptome data from expression sequence tag and RNA sequencing using next-generation sequencing have been made available and a genome project is in progress. However, the fact remains that the allo-tetraploid genome of *X. laevis* makes it difficult to perform genetic approaches, as more than 80 % of the genes are thought to be duplicated through polyploidization (Hellsten et al. 2007; Uno et al. 2013). Moreover, this species has a long maturation time of at least 2 years. To overcome these problems in *X. laevis*, a related species, *Xenopus* (*Silurana*) *tropicalis*, has recently been introduced into biological research (Harland and Grainger 2011). Its genome size is approximately 1.7 Gb and the sequence has already been elucidated (Hellsten et al. 2010). More notably, *X. tropicalis* has a diploid genome, and its generation time is about 1 year and consequently shorter than that of *X. laevis*. Therefore, this species could play a key role in amphibian research.

Urodeles including *Ambystoma mexicanum* (axolotl) and *Cynops pyrrhogaster* (newt), have a remarkable capability to regenerate lost body parts such as limbs, optical tissues, brain, spinal cord, intestine, and heart throughout their lives (reviewed in Brockes and Kumar 2002; Lepperdinger et al. 2008; Agata and Inoue 2012). Therefore, they have been used as unique experimental models for more than a century in regeneration biology. Thus far, axolotls have mainly been used for gene manipulation, including transgenic techniques, rather than newts. The conventional newts are not suitable as model animals for gene manipulation, because they have associated difficulties in rearing, feeding, and breeding. For these reasons, we have introduced the Iberian ribbed newt (*Pleurodeles waltl*) as a new animal model among newts. *P. waltl* has a short maturation time of less than 1 year and its eggs can be obtained throughout the year. We have established that the breeding system and transgenic technique in *P. waltl* are comparable to those in

Fig. 9.1 Model amphibians in this chapter. Two anurans (*Xenopus laevis* and *Xenopus (Sirulana) tropicalis*) and three urodeles (*Ambystoma mexicanum*, *Cynops pyrrhogaster*, and *Pleurodeles waltl*) are shown herein. The photographs were kindly provided by Dr. H. Hanada and Ms. Chiyo Nanba in Institute for Amphibian Biology, Graduate School of Science, Hiroshima University

Xenopus and *A. mexicanum* (Hayashi et al. 2013). Therefore, this newt has recently received a lot of attention in regeneration biology.

In amphibian studies, researchers often introduce mRNAs and antisense oligonucleotides for genes of interest into fertilized eggs to analyze their functions in vivo. Gain- and loss-of-function analyses using these techniques are convenient and effective in early embryogenesis. However, the effects are lost over time during development. In particular, these techniques cannot be applied to post-embryogenesis studies on processes such as metamorphosis, regeneration, and reproduction. Reliable and convenient transgenic techniques permit us to overexpress constitutively active and dominant-negative mutants of genes of interest to evaluate their functions in vivo in amphibians; however, the results may involve artifacts and lead us to erroneous conclusions. Therefore, gene targeting techniques for precise analyses of gene functions are extremely desirable. Recent progress in genome editing using site-specific nucleases such as zinc finger nucleases (ZFNs), transcription activator-like effector nucleases (TALENs), and clustered regularly interspaced short palindromic repeats (CRISPR)/CRISPR-associated (Cas) has made it possible to readily perform targeted gene knockout and knockin in various organisms (reviewed by Urnov et al. 2010; Gaj et al. 2013; Sander and Joung 2014; Hsu et al. 2014; Carroll 2014). In this chapter, we review recent studies involving genome editing with these nucleases in amphibians and discuss their possibilities in future biological research.

9.2 A Strategy for Genome Editing Using Site-Specific Nucleases in Amphibians

A strategy for in vivo gene targeting using ZFNs, TALENs, and CRISPR/Cas9 in amphibians is illustrated in Fig. 9.2. A great advantage of using amphibians in genome editing is that large amounts of mRNAs encoding site-specific nucleases can be directly injected into their fertilized eggs at the one- or two-cell stage. The translated site-specific nucleases introduce DNA double-strand break (DSB) at their target sites in the genome during the cleavage stage. According to our previous data using TALENs, somatic mutations may already be introduced at the morula stage when nuclease mRNAs are injected at the one-cell stage in *X. laevis* (Sakane et al. 2014). The target site is resected and then alternatively, but mainly, repaired by non-homologous end-joining (NHEJ) repair. In the case of gene knockout, insertions and deletions are induced by error-prone NHEJ repair at the target sites, and consequently the coding regions of the targeted genes are often inactivated by frameshift mutations. In amphibians, site-specific nucleases frequently alter target sites in both alleles with high efficiency, and therefore it is likely to be difficult to obtain Founder 1 (F1) progeny of genes that have crucial roles in early development. If knockout in F0 is the only objective, phenotypes may be sufficient to analyze the functions of target genes in amphibians, as described hereafter. For genotyping, site-specific nuclease-mediated mutations can be easily and reliably detected using the following methods: heteroduplex mobility assay (HMA), PCR-based restriction fragment length polymorphism (RFLP) analysis, and DNA sequencing (Ota et al. 2013; Suzuki et al. 2013; Nakagawa et al. 2014).

For gene knockin, single-stranded oligonucleotides (ssODNs) and targeting donor vectors carrying left and right homology arms can be introduced concomitantly with nuclease mRNAs targeting particular genomic loci. In this case, the ssODNs and donor vectors are thought to become integrated via homology-directed repair (HDR). ssODNs may be added to functional sequences such as loxP sites for Cre-mediated recombination leading to gene excision and short tag sequences such

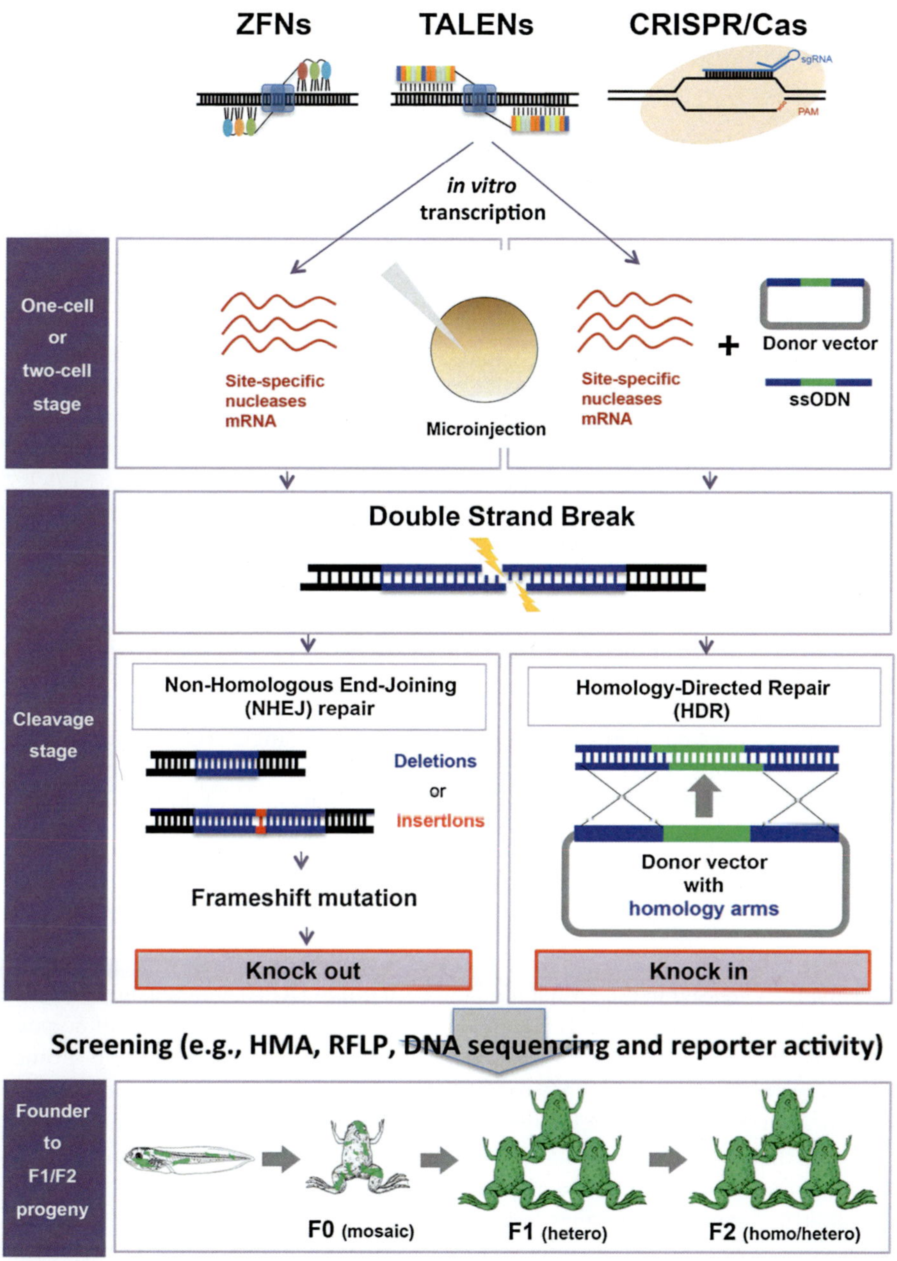

Fig. 9.2 Strategy for genome editing using site-specific nucleases in amphibians

as FLAG and hemagglutinin for in vivo tagging. If mutations and integration are induced at the target sites in germ cells, these genetic modifications are inherited by the F1 progeny.

9.3 Targeted Gene Disruption Using ZFNs in *Xenopus*

ZFNs were initially introduced into genome editing research in various organisms (reviewed by Urnov et al. 2010). ZFNs have the characteristic of Cys2/His2-type zinc finger (ZF) domains. A single ZF domain recognizes a 5′-G/A/CNN-3′ triplet sequence (N means any nucleotide), and a tandem array of 3–6 ZF domains is usually fused to the nuclease domain of the type II restriction enzyme *FokI*. ZFNs function as a pair, because the *FokI* nuclease domains must dimerize to cleave DNA. For instance, a pair of three-finger ZFNs that each recognize two 9-bp adjacent reverse target sites cleaves the spacer sequence between the two sites. To improve the specificity and decrease off-target effects, some *FokI* variants have been reported, such as EL/KK, ELD/KKR, and *Sharkey* mutations (Miller et al. 2007; Szczepek et al. 2007; Guo et al. 2010; Doyon et al. 2011). Detailed information about ZFNs is provided in Chap. 1 of this book (written by Ochiai and Yamamoto). Only two reports regarding ZFN-mediated mutations in amphibians have been published to date.

The first study about genome editing using site-specific nucleases in amphibians was reported by Young et al. (2011). They demonstrated that enhanced green fluorescence protein (egfp) gene is disrupted by injecting a pair of ZFN mRNAs into fertilized eggs derived from a transgenic line carrying a single copy of the reporter gene in *X. tropicalis*. The F0 phenotypes showed mosaic loss of EGFP fluorescence with somatic mutations of the target site. Targeted mutagenesis of *noggin*, an antagonist of bone morphogenic proteins (BMPs), was also successful and was inherited through the germline to the F1 progeny. They also reported that obligate heterodimeric *FokI* (EL/KK) reduces toxicity compared with homodimeric type.

Later, Nakajima et al. (2012) reported targeted disruption of *tyrosinase* (*tyr*) in *X. tropicalis* using ZFNs. The *tyr* gene encodes a melanin synthesis enzyme that is exclusively expressed in the retinal pigment epithelium (RPE) and melanophores in amphibians (Kumasaka et al. 2003), and its disruption results in albino phenotypes. Therefore, the efficacy of ZFNs can be readily evaluated by observing albino phenotypes without staining. They used custom-made ZFNs purchased from Sigma-Aldrich. The phenotypes were highly mosaic for albino phenotypes in F0 with various somatic mutations, but some full albino F1 progeny were obtained from mating of the mosaic F0 parents.

Researchers can introduce alterations at target genomic sites using ZFNs and obtain F1 progeny from F0 founders with germline mutations in the case of *X. tropicalis*, because of the short generation time. Thus, ZFNs are available for genome editing in *X. tropicalis*. However, the construction of highly active ZFNs may be a difficult task in a general laboratory, because the de novo assembly of ZF domains is an intricate process and the activity of ZF arrays is often affected by context-dependency (see Chap. 1).

9.4 Targeted Gene Disruption Using TALENs in *Xenopus*

In 2011, TALENs have been introduced into genome editing research as an effective site-specific nuclease and are based on TALE protein derived from the *Xanthomonas* genus, a bacterial plant pathogen (Miller et al. 2011). TALENs have the characteristics of high modularity in DNA-binding motif containing repeat variable di-residues (RVDs) that recognizes a single nucleotide, referred to as a TALE repeat. Researchers can readily construct an array of TALE repeats directed toward almost any sequence using the Golden Gate method (Cermak et al. 2011; Sakuma et al. 2013a). An array of TALE repeats targeting the sequence of interest is fused to TALE scaffold and heterodimeric or homodimeric types of the *FokI* nuclease domain described above. Although there are many reports regarding TALE scaffolds, truncated scaffold carrying N terminal domain +136 and C terminal domain +63 (referred to as +136/+63) is widely used and exhibits high activity in various organisms (Miller et al. 2011; Sander et al. 2011; Reyon et al. 2012). As with ZFNs, a single pair of TALENs dimerizes and then cleaves a spacer sequence between their recognition sequences. TALENs have several advantages over ZFNs: specificity, efficiency, and easy construction. Detailed information about TALENs is provided in Chap. 1 of this book (written by Ochiai and Yamamoto).

Ishibashi et al. (2012) initially reported gene disruption of *tyr* in *X. tropicalis* using TALENs. They used a single pair of TALENs targeting exon 1 of *tyr* that were purchased from Cellectis. They subcloned these TALENs into pCS2+, a commonly available in vitro transcription vector for *Xenopus* researchers (Turner and Weintraub 1994). In vitro-transcribed mRNAs were injected into each blastomere at the two-cell stage. Surprisingly, introduction of the TALENs targeting *tyr* resulted in >90 % partial or full albino phenotypes in F0 embryos. DNA sequencing showed alterations of the target site with high efficiency (~70 %) in each phenotype. Furthermore, the mutations of *tyr* were inherited through the germline to the F1 progeny showing full albino phenotypes, suggesting that TALENs represent a highly effective tool to disrupt genes of interest in reverse genetics research using *X. tropicalis*.

Soon after, Lei et al. (2012) reported a more practical study of genome editing using TALENs in *X. tropicalis*. They selected a highly active +136 (Δ152)/+63 TALE scaffold and subcloned it into pCS2+ used as a destination and in vitro transcription vector. Obligate heterodimer mutants of the ELD/KKR *FokI* nuclease domain were selected to decrease off-target effects (Doyon et al. 2011). They examined *ptf1a/p48*, *ets1*, *noggin*, *hhex*, *vpp1*, *foxd3*, *sox9*, and *grp78/bip*, all of which exhibited somatic mutations with high efficiencies. Among them, mutations of *ets1* and *ptf1a/p48* were inherited by the F1 progeny. They also described much lower toxicities of the TALENs targeting *noggin* and *ptf1a/p48* than those of ZFNs.

X. laevis has an allo-tetraploid genome and >80 % of its genes are thought to be duplicated as homeologs by polyploidization (Hellsten et al. 2007; Uno et al. 2013). Individual homeologs have often functional redundancy. Therefore, to precisely analyze the functions of genes, simultaneous disruption of both homeologs is needed. Although Lei et al. (2012) also reported one example of *ets1a*-targeted mutagenesis, we reported a more practical study of genome editing using TALENs

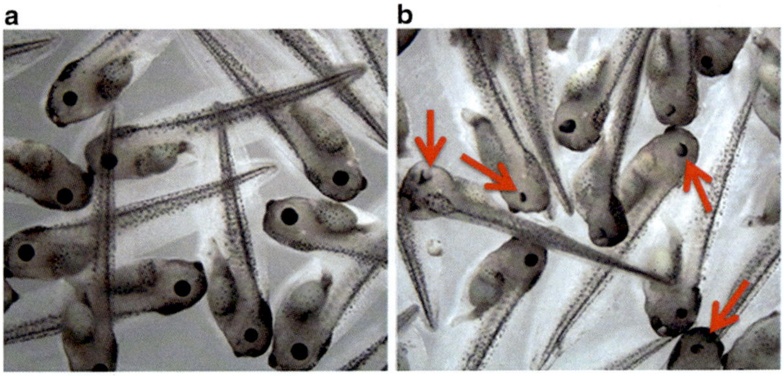

Fig. 9.3 Targeted mutagenesis of the *pax6* gene in F0 embryos of *Xenopus laevis*. Phenotypes of *pax6* TALEN-injected embryos are shown: (**a**) TALEN-R, embryos injected with 1,100 pg of right *pax6* TALEN mRNA; (**b**) TALEN-R/L, embryos injected with 550 pg each of right and left *pax6* TALEN mRNAs. The injected embryos were reared to the hatching stage. Deformation of the eyes was observed in the TALEN-R/L-injected embryos (*arrows* in **b**). Note that biallelic and homeologous disruptions of *pax6* were introduced in these phenotypes (Suzuki et al. 2013)

in *X. laevis* (Suzuki et al. 2013). To address the issue regarding homeologs, we designed and used a single pair of TALENs targeting conserved sequences between homeologs, *tyr* and *pax6*, which are involved in pigmentation and eye formation, respectively. We used the +153/+47 TALE scaffold containing a homodimer of the *FokI* nuclease domain, because this combination showed low toxicity in *X. laevis* (Mussolino et al. 2011; Sakuma et al. 2013a). In vitro-transcribed mRNAs encoding TALENs were injected into fertilized eggs at the one-cell stage. Introduction of the *tyr* TALEN mRNAs resulted in drastic loss of pigmentation in the RPE and melanophores with high frequencies of the expected phenotypes. Similarly, the *pax6* TALENs led to deformed eyes in the injected embryos consistent with a previous report of knock-down phenotypes by morpholino antisense oligonucleotides (MO) (Fig. 9.3; Rungger-Brändle et al. 2010). Although there were at least four target sites in the *X. laevis* genome, the somatic mutation rates of *tyr* (*tyra* and *tyrb*) and *pax6* (*pax6a* and *pax6b*) homeologs were >50 % and >90 %, respectively, indicating that not only biallelic but also homeologous disruption of the target sites was induced by a single pair of TALENs recognizing the conserved sequences between each gene. The toxicity was very low (~10 %) even though >1,000 pg/egg of TALEN mRNAs was injected. These results demonstrate that targeted gene disruption by TALENs provides a robust tool to analyze gene functions in *X. laevis* F0 embryos. Full albino F1 progeny were obtained from mating of the mosaic F0 parents, and therefore mutations of *tyr* were definitely inherited through the germline into the F1 progeny (Fig. 9.4; our unpublished data).

Next, focusing on the repeating pattern of non-RVD variations in TALE repeats, we developed a highly active TALEN system, referred to as the Platinum TALEN system (Sakuma et al. 2013b). The activity of Platinum TALENs is markedly

Fig. 9.4 Germline transmission of *tyrosinase* mutations into the F1 progeny of *Xenopus laevis*. Mosaic albino F0 phenotypes were reared to fully-mature adults and F1 progeny were generated by brother–sister inbreeding of F0. Most of the eggs were albino. *Upper*: full-albino F1 progeny. *Lower*: wild-type F1 progeny. TALEN-mediated mutations of *tyrosinase* were definitely inherited through the germ line to the F1 progeny. The photograph was kindly provided by Prof. A. Kashiwagi and Dr. K. Kashiwagi in Institute for Amphibian Biology, Graduate School of Science, Hiroshima University

increased compared with the previous types, while maintaining low toxicity (Sakuma et al. 2013a). Therefore, we examined multiplex gene disruption using Platinum TALENs in *X. laevis* (Sakane et al. 2014). In the study, we attempted multiplex gene disruption of *tyr* homeologs (*tyra* and *tyrb*) and egfp by injection of two TALEN pairs into transgenic embryos carrying multiple copies of egfp. Surprisingly, and consistent with the high frequencies of both severe phenotypic traits (albinism and loss of GFP fluorescence), the frameshift mutation rates of the *tyr* homeologs and egfp reached 40–80 %. We also showed that two different pairs of TALENs could simultaneously introduce mutations in both homeologs encoding histone chaperones, *no29* and *npm3*, with high efficiency. The results of the study suggest that multiplex gene disruption using TALENs can be applied to analyze the functions of homeologous genes with redundancy in F0 *X. laevis* embryos. Our results also indicated that TALEN-mediated mutagenesis begins to occur at least in the morula stage, about 5 hours after injection. This result suggests that TALENs possess the potential to disrupt target genes at relatively early stages, even though four copies of the genes exist in the genome. Thus, such early introduction of targeted mutagenesis by TALENs allows us to perform loss-of-function analyses in F0 embryos during early embryogenesis.

It was reported that Goldy TALEN (Δ152 or +136/+63) with an obligate ELD/KKR *FokI* nuclease domain exhibited a high efficiency of gene mutations with low toxicity (Bedell et al. 2012). Nakajima and Yaoita (2013) evaluated four TALE scaffolds and *FokI* variants targeting *tyr* and showed efficacy of Goldy TALENs with ELD/KKR in *X. tropicalis*. We also examined various combinations of TALE repeats,

TALE scaffolds and *FokI* variants, and as a result, we believe that Platinum TALENs using +153/+47 scaffold with the homodimer type of *FokI* represent an excellent tool for genome editing in *X. laevis* (Sakuma et al. 2013b; our unpublished data). We herein emphasize that optimization of TALE repeats, TALE scaffolds and *FokI* variants for each species is needed for genome editing studies on amphibians.

9.5 Targeted Gene Disruption Using CRISPR/Cas9 in *Xenopus*

The CRISPR/Cas9 system is the most recent technique in genome editing using site-specific nucleases, and is based on an adaptive immune system in bacteria and archaea (Jinek et al. 2012; Mali et al. 2013; Cong et al. 2013). The CRISPR/Cas9 technique is explosively expanding in life science due to its simplicity. The Cas9 endonuclease is guided by a single guide RNA (sgRNA), which is a chimeric RNA of trans-activating CRISPR RNA (tracrRNA) and CRISPR RNA (crRNA) containing 20 nucleotides of the target sequence. 5′-NGG-3′ is known as protospacer adjacent motif (PAM) for the recognition of Cas9 derived from *Streptococcus pyogenes* (*sp*Cas9). Sequence at the 3′ end of target site must be followed by PAM. Cas9 acts as a monomer and sgRNA/Cas9 complex introduces DSB into the target site, therefore, this system is defined as RNA-guided nuclease. Detailed information about the CRISPR/Cas9 system is provided in Chap. 2 of this book (written by Sakuma and Yamamoto).

Blitz et al. (2013) and Nakayama et al. (2013) initially reported gene disruption by the CRISPR/Cas9 system in amphibians. Both groups demonstrated *tyr* disruption by coinjection of Cas9 mRNA and an sgRNA targeting exon 1 of *tyr* into fertilized eggs at the one- or two-cell stage of *X. tropicalis*. The sgRNA/Cas9-injected embryos showed various degrees of albino phenotypes, including almost full lack of pigmentation in the RPE and melanophores. Nakayama et al. (2013) also reported gene disruption of *six3* involved in eye formation as a key transcription factor. They designed two sgRNAs targeting its coding and promoter regions. Each sgRNA/Cas9-injected embryo showed severe deformation of the eyes and anterior head region consistent with a previous report of MO knock-down phenotypes in *X. laevis* (Gestri et al. 2005). Moreover, the sgRNA/Cas9-injected embryos could be rescued by injection of an mRNA encoding *X. laevis six3*. Such rescue experiments are required for precise analyses of gene functions in future amphibian genome editing research using the CRISPR/Cas9 system.

Soon after, Guo et al. (2014) reported a practical and robust study on CRISPR/Cas9-mediated genome editing in *X. tropicalis*. They examined the efficiencies and toxicities of 18 targeted loci including *tyr* with various doses of Cas9 mRNA and sgRNAs. High doses of Cas9 mRNA and sgRNAs caused high toxicity in the injected embryos. Under the optimized amounts of Cas9 mRNA and sgRNAs for each gene, high mutation rates and severe phenotypes were observed. For example, 61 % of *tyr* sgRNA/Cas9-injected embryos exhibited severe phenotypes lacking pigmentation in the RPE and melanophores, including a few full albino phenotypes. Targeted mutations of *tyr*, *elastase-T1*, and *elastase-T2* were inherited through the germline to the F1 progeny with high efficiencies (40–100 %). In some cases, sgRNA/Cas9 could not introduce any gene alterations into the target sites, and therefore researchers need to design and prepare several sgRNAs for each gene. In the CRISPR/Cas9 system, multiplex gene disruption is one of the attractive advantages (Mali et al. 2013; Cong et al. 2013). When multiplex sgRNAs for different loci were concomitantly introduced with Cas9 mRNA into eggs, gene alterations of all targets could be achieved in vivo (Wang et al. 2013; Jao et al. 2013). In fact, Guo et al. (2014) also succeeded in simultaneous gene disruption of *grp78* and *elastase-T1* in *X. tropicalis*. These multiple mutations were also inherited through the germline to the F1 progeny. Furthermore, multiplex gene disruption using CRISPR/Cas9 is effective against not only homeologs but also paralogs such as the *hox* cluster genes that have functional redundancy in *X. laevis*.

Off-target effects are a matter of concern when researchers perform genome editing using site-specific nucleases. Although the CRISPR/Cas9 system is a very convenient method, some reports have raised the issue of off-target effects using this system (Fu et al. 2013; Cradick et al. 2013). When the CRISPR/Cas9 system is used in vivo, researchers need to pay attention to unintended off-target effects and perform some control experiments. For example, mismatch sgRNAs should be introduced as a negative control. To confirm knockout experiments, some sgRNAs for other sites in a target gene also need to be performed. Rescue experiments using injection of an mRNA encoding the target gene and a transgenic technique would be useful. Since *X. tropicalis* genome data are well-annotated and available, we can carefully design sgRNAs and predict candidates for off-target sequences in the genome (Hellsten et al. 2010). However, genome data of *X. laevis* are not yet available, and its size is much larger than that of *X. tropicalis*. Therefore, researchers require further consideration of off-target effects in this species.

9.6 Targeted Gene Disruption Using TALENs and CRISPR/Cas9 in Urodeles

Thus far, there are no reports on genome editing using ZFNs in urodeles. Using *P. waltl* newts, we first attempted to disrupt the *tyr* gene to examine whether TALENs can induce targeted mutagenesis (Hayashi et al. 2014). As shown in Fig. 9.5, no melanin-pigmented cells were observed in animals injected with the right and left TALEN mRNAs. In contrast, injection with only the right TALEN mRNA as a negative control produced the wild-type phenotype (Fig. 9.5a, white arrow). To confirm disruption of the *tyr* gene in the albino animals, we sequenced genomic PCR products of the target region from three albino larvae, and the results indicated 6–11-bp deletions in all clones examined (Fig. 9.5b).

Conditional knockout animals are required to avoid embryonic lethality or malformations in post-embryonic events such as metamorphosis and regeneration

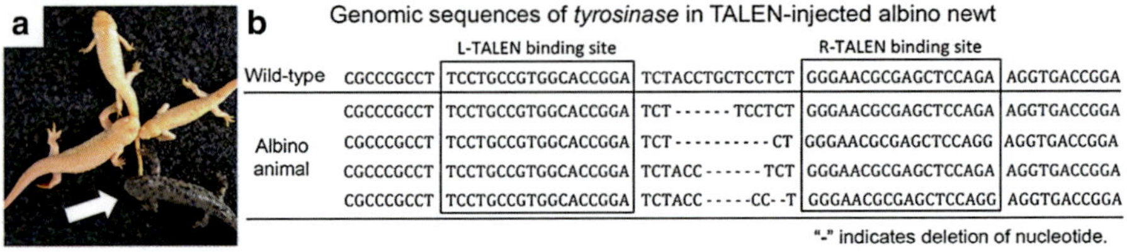

Fig. 9.5 Targeted mutagenesis of *tyrosinase* in F0 *Pleurodeles waltl.* (**a**) Albino phenotypes in F0 adult newts targeting *tyrosinase*. Fertilized eggs were injected with 100 pg each of R- and L-TALEN mRNAs (total: 200 pg). The grown-up newts showed almost complete albino phenotypes. The *white arrow* indicates an animal injected with the R-TALEN only (200 pg) that showed the wild-type body color. (**b**) Genomic sequences observed in *tyrosinase* knockout larvae. The wild-type sequence is shown at the *top*

studies using tadpoles and adults. Therefore, we tried to introduce two loxP sites flanking an important exon of the target gene. When these newts are crossed with inducible Cre recombinase lines, the target genes can be conditionally disrupted (Whited et al. 2012; Khattak et al. 2013; our unpublished data). An ssODN containing a loxP site and an EcoRI site was injected with TALEN mRNAs targeting the *tyr* locus (Fig. 9.6a). This ssODN has 26-mer homology arms for insertion into the *tyr* locus via HDR. The targeted insertion was detected by digestion of genomic PCR products with EcoRI, as shown in Fig. 9.6b. Genomic sequencing demonstrated that precise insertion of the ssODN had occurred in ~10 % of the injected animals (Fig. 9.6c). This result suggests that conditional knockout newts will be available in the near future.

In addition to TALENs, Flowers et al. (2014) reported that CRISPR/Cas9 also works efficiently in the axolotl. They demonstrated the disruption of egfp in transgenic animals carrying a single copy reporter gene. sgRNA/Cas9 mRNA-injected embryos displayed drastic loss of systemic EGFP fluorescence. Next, they successfully produced developmental phenotypes by disruption of *brachyury* and *bambi* orthologs, which represent a key transcription factor in mesoderm formation and an antagonist of BMP signaling, respectively. Notably, *brachyury* sgRNA/Cas9 mRNA-injected embryos showed clear phenotypes that resulted in an anteroposterior axis defect and extensive tail shortening with high mutation rates of the target site (close to 100 %). Their report confirms the practicality of the CRISPR/Cas9 system for gene knockout in urodele research.

These findings demonstrate that genome editing using TALENs and CRISPR/Cas9 represents a powerful reverse genetics tool to generate not only knockout but also knockin animals in urodeles. However, there is an issue concerning the use of TALENs and CRISPR/Cas9 in urodeles, namely unexpected off-target effects due to the size of the genome. Urodeles have huge genomes exceeding 10 Gb that include highly repetitive sequences. Therefore, their whole genomes have not been sequenced and made available. We prefer to use TALENs for precise functional analyses of genes in urodeles, because we assume that the off-targeting frequency of TALENs will be lower than that of CRISPR/Cas9. However, the fact remains that

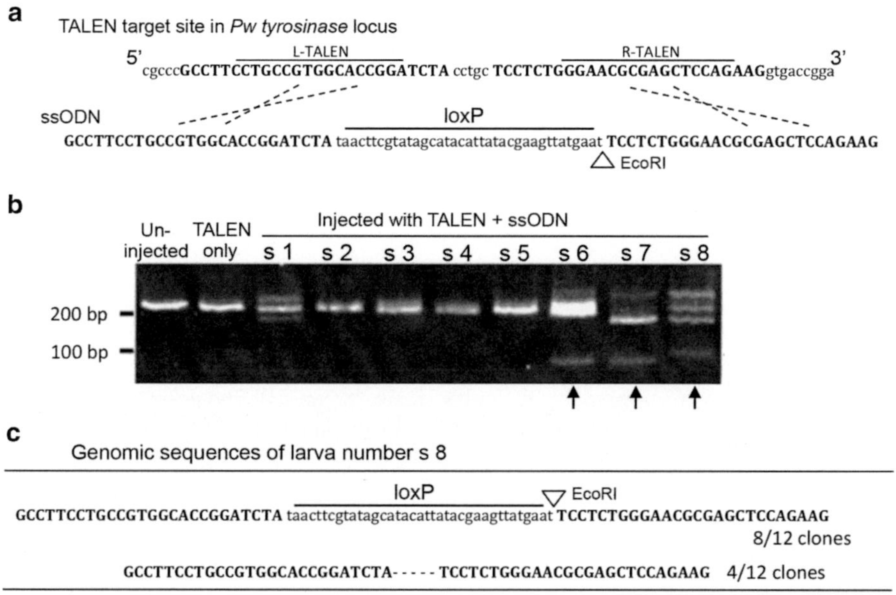

Fig. 9.6 Knockin of a single-stranded oligonucleotide (ssODN) into the *tyrosinase* locus in F0 *Pleurodeles waltl*. (**a**) Schematic diagrams of the *P. waltl tyrosinase* locus (*upper*) and ssODN (*lower*). The homology arms are shown in *capital letters*. The *arrowhead* indicates the EcoRI digestion site. The *bar* indicates the loxP site (34 bases). (**b**) Gel image of genomic PCR products from TALEN-injected larvae digested with EcoRI. The *numbers s1–s8* represent individual larvae. The products of *s6, s7, and s8* were digested with EcoRI (*arrows*). (**c**) Genomic sequences from larva number s8 shown in (**b**). Exact insertion of the loxP and EcoRI sites was confirmed in 8 of 12 clones (*upper*), while 4 of 12 clones showed five base deletions (*lower*). This figure is modified from Hayashi et al. (2014)

it is relatively easy to use the CRISPR/Cas9 system. Hereafter, TALENs or CRISPR/Cas9 will be chosen according to the experimental designs and objectives. Establishment of genome editing in urodeles will promote our understanding of the mechanisms of regeneration. In addition, urodeles would represent valuable experimental models for studies involving development, stem cells, and reprogramming because of their unique biological properties.

Concluding Remarks

Researchers have been inspired by the drastic morphological changes that occur during embryogenesis, metamorphosis, and regeneration in amphibians. Thus far, amphibians have a long history of research in developmental and cell biology, and have greatly contributed to the discovery of the general principles of biology through their excellent properties as model animals.

Numerous fertilized eggs at the single-cell stage can be obtained when needed, and their eggs and embryos successfully endure surgical manipulation and transplantation. Furthermore, large amounts of mRNAs and oligonucleotides for analyses of gene function can be introduced into eggs and oocytes by injection. In addition to these properties, gene knockout and knockin using genome editing techniques have become available as indispensable tools. These techniques have long been desired for amphibian research, but could not previously be performed because of the absence of effective tools. The emergence of site-specific nucleases brings innovation to amphibian research and will further enhance the value of these species as model animals in biology.

In this chapter, we have introduced recent studies using genome editing in the amphibians *X. tropicalis*, *X. laevis*, *P. waltl*, and *A. mexicanum*, all of which are highly active model animals in developmental biology and regeneration biology. Site-specific nucleases, especially TALENs and CRISPR/Cas9, exhibit quite high efficiency of targeted gene alterations in all of these species. Most importantly, such high efficiency of site-specific nucleases in amphibians can introduce biallelically targeted gene disruption in F0 animals, and this allows us to save time in generating mutant offspring. This will be an invaluable advantage for most amphibians, because the production of their mutant offspring requires long generation times in comparison with other species such as rodents and fishes. As described above, *X. tropicalis* and *P. waltl* have relatively short maturation times of about 1 year, and we can therefore analyze gene functions using their F1/F2 offspring carrying homo- and hetero-mutant alleles within a realistic time frame.

Moreover, it is suggested that fully-fledged gene targeting, such as gene knockin, is feasible using TALENs and CRISPR/Cas9 in amphibians. In fact, we have demonstrated precise TALEN-mediated insertion of a loxP site into a particular target locus of the *P. waltl* genome using an ssODN. More advanced manipulation of gene expression in vivo, such as conditional knockout, exchange of promoter sequence, in vivo protein tagging and reporter knockin, is no longer a dream in amphibians. Unfortunately, gene knockout has never previously been accomplished in *C. pyrrhogaster* (Japanese fire belly newt) and *Notophthalmus viridescens* (Red-spotted newt), which are traditional and attractive animal for regeneration studies. We expect that genome editing in these newts will also be successful in the near future.

As mentioned above, site-specific nuclease-mediated genome editing makes it possible to perform reverse genetic studies that were previously difficult or impossible in amphibians. The research fields and applications of frogs and newts will become increasingly expanded by this technique. In particular, we are interested in post-embryogenesis events such as metamorphosis and regeneration in amphibians. The genome editing approach would be a powerful tool to elucidate the molecular mechanisms underlying these phenomena. We would like many researchers to join in these research fields with novel ideas that can be addressed by this technique.

Acknowledgements We gratefully thank Profs. Takashi Yamamoto, Takashi Takeuchi, and Akihiko Kashiwagi, Drs. Tetsushi Sakuma and Keiko Kashiwagi, Mr. Yuto Sakane, and Ms. Miyuki Suzuki for their valuable help and advice. This work was supported by a Grant-in-Aid for Young Scientists (B) to T. H. (25840086) and Grants-in-Aid for Scientific Research on Innovative Areas to K.T.S. (25124708) and T.H. (25124706) from the Ministry of Education, Science, Sports and Culture of Japan.

References

Agata K, Inoue T (2012) Survey of the differences between regenerative and non-regenerative animals. Dev Growth Differ 54:143–152

Bedell VM, Wang Y, Campbell JM, Poshusta TL, Starker CG, Krug RG 2nd, Tan W, Penheiter SG, Ma AC, Leung AY, Fahrenkrug SC, Carlson DF, Voytas DF, Clark KJ, Essner JJ, Ekker SC (2012) In vivo genome editing using a high-efficiency TALEN system. Nature 491:114–118

Blitz IL, Biesinger J, Xie X, Cho KW (2013) Biallelic genome modification in F(0) *Xenopus tropicalis* embryos using the CRISPR/Cas system. Genesis 51:827–834

Brockes JP, Kumar A (2002) Plasticity and reprogramming of differentiated cells in amphibian regeneration. Nat Rev Mol Cell Biol 3:566–574

Carroll D (2014) Genome engineering with targetable nucleases. Annu Rev Biochem 83:409–439

Cermak T, Doyle EL, Christian M, Wang L, Zhang Y, Schmidt C, Baller JA, Somia NV, Bogdanove AJ, Voytas DF (2011) Efficient design and assembly of custom TALEN and other TAL effector-based constructs for DNA targeting. Nucleic Acids Res 39:e82

Cong L, Ran FA, Cox D, Lin S, Barretto R, Habib N, Hsu PD, Wu X, Jiang W, Marraffini LA, Zhang F (2013) Multiplex genome engineering using CRISPR/Cas systems. Science 339:819–823

Cradick TJ, Fine EJ, Antico CJ, Bao G (2013) CRISPR/Cas9 systems targeting beta-globin and CCR5 genes have substantial off-target activity. Nucleic Acids Res 41:9584–9592

Doyon Y, Vo TD, Mendel MC, Greenberg SG, Wang J, Xia DF, Miller JC, Urnov FD, Gregory PD, Holmes MC (2011) Enhancing zinc-finger-nuclease activity with improved obligate heterodimeric architectures. Nat Methods 8:74–79

Flowers GP, Timberlake AT, McLean KC, Monaghan JR, Crews CM (2014) Highly efficient targeted mutagenesis in axolotl using Cas9 RNA-guided nuclease. Development 141:2165–2171

Fu Y, Foden JA, Khayter C, Maeder ML, Reyon D, Joung JK, Sander JD (2013) High-frequency off-target mutagenesis induced by CRISPR-Cas nucleases in human cells. Nat Biotechnol 31:822–826

Gaj T, Gersbach CA, Barbas CF 3rd (2013) ZFN, TALEN, and CRISPR/Cas-based methods for genome engineering. Trends Biotechnol 31:397–405

Gestri G, Carl M, Appolloni I, Wilson SW, Barsacchi G, Andreazzoli M (2005) Six3 functions in anterior neural plate specification by promoting cell proliferation and inhibiting Bmp4 expression. Development 132:2401–2413

Guo J, Gaj T, Barbas CF 3rd (2010) Directed evolution of an enhanced and highly efficient FokI cleavage domain for zinc finger nucleases. J Mol Biol 400:96–107

Guo X, Zhang T, Hu Z, Zhang Y, Shi Z, Wang Q, Cui Y, Wang F, Zhao H, Chen Y (2014) Efficient RNA/Cas9-mediated genome editing in *Xenopus tropicalis*. Development 141:707–714

Harland RM, Grainger RM (2011) *Xenopus* research: metamorphosed by genetics and genomics. Trends Genet 27:507–515

Hayashi T, Yokotani N, Tane S, Matsumoto A, Myouga A, Okamoto M, Takeuchi T (2013) Molecular genetic system for regenerative studies using newts. Dev Growth Differ 55:229–236

Hayashi T, Sakamoto K, Sakuma T, Yokotani N, Inoue T, Kawaguchi E, Agata K, Yamamoto T, Takeuchi T (2014) Transcription activator-like effector nucleases efficiently disrupt the target gene in Iberian ribbed newts (*Pleurodeles waltl*), an experimental model animal for regeneration. Dev Growth Differ 56:115–121

Hellsten U, Khokha MK, Grammer TC, Harland RM, Richardson P, Rokhsar DS (2007) Accelerated gene evolution and subfunctionalization in the pseudotetraploid frog *Xenopus laevis*. BMC Biol 5:31

Hellsten U, Harland RM, Gilchrist MJ, Hendrix D, Jurka J, Kapitonov V, Ovcharenko I, Putnam NH, Shu S, Taher L, Blitz IL, Blumberg B, Dichmann DS, Dubchak I, Amaya E, Detter JC, Fletcher R, Gerhard DS, Goodstein D, Graves T, Grigoriev IV, Grimwood J, Kawashima T, Lindquist E, Lucas SM, Mead PE, Mitros T, Ogino H, Ohta Y, Poliakov AV, Pollet N, Robert J, Salamov A, Sater AK, Schmutz J, Terry A, Vize PD, Warren WC, Wells D, Wills A, Wilson RK, Zimmerman LB, Zorn AM, Grainger R, Grammer T, Khokha MK, Richardson PM, Rokhsar DS (2010) The genome of the Western clawed frog *Xenopus tropicalis*. Science 328:633–636

Hsu PD, Lander ES, Zhang F (2014) Development and applications of CRISPR-Cas9 for genome engineering. Cell 157:1262–1278

Ishibashi S, Cliffe R, Amaya E (2012) Highly efficient bi-allelic mutation rates using TALENs in *Xenopus tropicalis*. Biol Open 1:1273–1276

Jao LE, Wente SR, Chen W (2013) Efficient multiplex biallelic zebrafish genome editing using a CRISPR nuclease system. Proc Natl Acad Sci U S A 110:13904–13909

Jinek M, Chylinski K, Fonfara I, Hauer M, Doudna JA, Charpentier E (2012) A programmable dual-RNA-guided DNA endonuclease in adaptive bacterial immunity. Science 337:816–821

Khattak S, Schuez M, Richter T, Knapp D, Haigo SL, Sandoval-Guzman T, Hradlikova K, Duemmler A, Kerney R, Tanaka EM (2013) Germline transgenic methods for tracking cells and testing gene function during regeneration in the axolotl. Stem Cell Reports 1:90–103

Kumasaka M, Sato S, Yajima I, Yamamoto H (2003) Isolation and developmental expression of tyrosinase family genes in *Xenopus laevis*. Pigment Cell Res 16:455–462

Lei Y, Guo X, Liu Y, Cao Y, Deng Y, Chen X, Cheng CH, Dawid IB, Chen Y, Zhao H (2012) Efficient targeted gene disruption in *Xenopus* embryos using engineered transcription activator-like effector nucleases (TALENs). Proc Natl Acad Sci U S A 109:17484–17489

Lepperdinger G, Brunauer R, Gassner R, Jamnig A, Kloss F, Laschober GT (2008) Changes of the functional capacity of mesenchymal stem cells due to aging or age-associated disease—implications for clinical applications and donor recruitment. Transfus Med Hemother 35:299–305

Mali P, Yang L, Esvelt KM, Aach J, Guell M, DiCarlo JE, Norville JE, Church GM (2013) RNA-guided human genome engineering via Cas9. Science 339:823–826

Miller JC, Holmes MC, Wang J, Guschin DY, Lee YL, Rupniewski I, Beausejour CM, Waite AJ, Wang NS, Kim KA, Gregory PD, Pabo CO, Rebar EJ (2007) An improved zinc-finger nuclease architecture for highly specific genome editing. Nat Biotechnol 25:778–785

Miller JC, Tan S, Qiao G, Barlow KA, Wang J, Xia DF, Meng X, Paschon DE, Leung E, Hinkley SJ, Dulay GP, Hua KL, Ankoudinova I, Cost GJ, Urnov FD, Zhang HS, Holmes MC, Zhang L, Gregory PD, Rebar EJ (2011) A TALE nuclease architecture for efficient genome editing. Nat Biotechnol 29:143–148

Mussolino C, Morbitzer R, Lutge F, Dannemann N, Lahaye T, Cathomen T (2011) A novel TALE nuclease scaffold enables high genome editing activity in combination with low toxicity. Nucleic Acids Res 39:9283–9293

Nakagawa Y, Yamamoto T, Suzuki K, Araki K, Takeda N, Ohmuraya M, Sakuma T (2014) Screening methods to identify TALEN-mediated knockout mice. Exp Anim 63:79–84

Nakajima K, Yaoita Y (2013) Comparison of TALEN scaffolds in *Xenopus tropicalis*. Biol Open 2:1364–1370

Nakajima K, Nakajima T, Takase M, Yaoita Y (2012) Generation of albino *Xenopus tropicalis* using zinc-finger nucleases. Dev Growth Differ 54:777–784

Nakayama T, Fish MB, Fisher M, Oomen-Hajagos J, Thomsen GH, Grainger RM (2013) Simple and efficient CRISPR/Cas9-mediated targeted mutagenesis in *Xenopus tropicalis*. Genesis 51:835–843

Ota S, Hisano Y, Muraki M, Hoshijima K, Dahlem TJ, Grunwald DJ, Okada Y, Kawahara A (2013) Efficient identification of TALEN-mediated genome modifications using heteroduplex mobility assays. Genes Cells 18:450–458

Reyon D, Tsai SQ, Khayter C, Foden JA, Sander JD, Joung JK (2012) FLASH assembly of TALENs for high-throughput genome editing. Nat Biotechnol 30:460–465

Rungger-Brändle E, Ripperger JA, Steiner K, Conti A, Stieger A, Soltanieh S, Rungger D (2010) Retinal patterning by Pax6-dependent cell adhesion molecules. Dev Neurobiol 70:764–780

Sakane Y, Sakuma T, Kashiwagi K, Kashiwagi A, Yamamoto T, Suzuki KT (2014) Targeted mutagenesis of multiple and paralogous genes in *Xenopus laevis* using two pairs of transcription activator-like effector nucleases. Dev Growth Differ 56:108–114

Sakuma T, Hosoi S, Woltjen K, Suzuki K, Kashiwagi K, Wada H, Ochiai H, Miyamoto T, Kawai N, Sasakura Y, Matsuura S, Okada Y, Kawahara A, Hayashi S, Yamamoto T (2013a) Efficient TALEN construction and evaluation methods for human cell and animal applications. Genes Cells 18:315–326

Sakuma T, Ochiai H, Kaneko T, Mashimo T, Tokumasu D, Sakane Y, Suzuki K, Miyamoto T, Sakamoto N, Matsuura S, Yamamoto T (2013b) Repeating pattern of non-RVD variations in DNA-binding modules enhances TALEN activity. Sci Rep 3:3379

Sander JD, Joung JK (2014) CRISPR-Cas systems for editing, regulating and targeting genomes. Nat Biotechnol 32:347–355

Sander JD, Cade L, Khayter C, Reyon D, Peterson RT, Joung JK, Yeh JR (2011) Targeted gene disruption in somatic zebrafish cells using engineered TALENs. Nat Biotechnol 29:697–698

Suzuki KT, Isoyama Y, Kashiwagi K, Sakuma T, Ochiai H, Sakamoto N, Furuno N, Kashiwagi A, Yamamoto T (2013) High efficiency TALENs enable F0 functional analysis by targeted gene disruption in *Xenopus laevis* embryos. Biol Open 2:448–452

Szczepek M, Brondani V, Buchel J, Serrano L, Segal DJ, Cathomen T (2007) Structure-based redesign of the dimerization interface reduces the toxicity of zinc-finger nucleases. Nat Biotechnol 25:786–793

Turner DL, Weintraub H (1994) Expression of achaete-scute homolog 3 in *Xenopus* embryos converts ectodermal cells to a neural fate. Genes Dev 8:1434–1447

Uno Y, Nishida C, Takagi C, Ueno N, Matsuda Y (2013) Homoeologous chromosomes of *Xenopus laevis* are highly conserved after whole-genome duplication. Heredity (Edinb) 111:430–436

Urnov FD, Rebar EJ, Holmes MC, Zhang HS, Gregory PD (2010) Genome editing with engineered zinc finger nucleases. Nat Rev Genet 11:636–646

Wang H, Yang H, Shivalila CS, Dawlaty MM, Cheng AW, Zhang F, Jaenisch R (2013) One-step generation of mice carrying mutations in multiple genes by CRISPR/Cas-mediated genome engineering. Cell 153:910–918

Whited JL, Lehoczky JA, Tabin CJ (2012) Inducible genetic system for the axolotl. Proc Natl Acad Sci U S A 109:13662–13667

Young JJ, Cherone JM, Doyon Y, Ankoudinova I, Faraji FM, Lee AH, Ngo C, Guschin DY, Paschon DE, Miller JC, Zhang L, Rebar EJ, Gregory PD, Urnov FD, Harland RM, Zeitler B (2011) Efficient targeted gene disruption in the soma and germ line of the frog *Xenopus tropicalis* using engineered zinc-finger nucleases. Proc Natl Acad Sci U S A 108:7052–7057

Chapter 10
Genome Editing in Mice Using CRISPR/Cas

Samantha A.M. Young, Mark Baker, and Masahito Ikawa

Abstract Mice have long been used as model organisms for investigating genetic features. The adaptation of clustered regularly interspaced palindromic repeats (CRISPR) and their associated nucleases (Cas) has increased our ability to utilise mice to further our knowledge of genetics as well as to create models of human disease conditions. Development of the CRISPR/Cas system is such that scientists can now create knockout/knockin mice in less than half the time previously required, as well as generate point mutations and insertion of small targeting sequences, allowing the finer detail of the genome to be examined. Current uses for the CRISPR/Cas system in mice also allows for deactivation of the cleavage domains of the Cas nuclease, permitting fusion of nickases or other activating/repressing effector proteins. This enables researchers to investigate timing of activation and localisation of a target locus and has even been used to correct genetic disease in the mouse. Mice are an essential part of the biological/biomedical research process and the CRISPR/Cas system has been and continues to be used in this organism with extremely promising results.

Keywords Gene repair • Gene tagging • Homology directed repair • Murine models • Non-Homologous end-joining

S.A.M. Young
Priority Research Centre in Reproductive Science, Discipline of Biological Sciences, Faculty of Science and IT, University of Newcastle, Callaghan, NSW, Australia

Animal Resource Center for Infectious Diseases, Research Institute for Microbial Diseases, Osaka University, Suita, Osaka, Japan

M. Baker
Priority Research Centre in Reproductive Science, Discipline of Biological Sciences, Faculty of Science and IT, University of Newcastle, Callaghan, NSW, Australia

M. Ikawa (✉)
Animal Resource Center for Infectious Diseases, Research Institute for Microbial Diseases, Osaka University, Suita, Osaka, Japan
e-mail: ikawa@biken.osaka-u.ac.jp

T. Yamamoto (ed.), *Targeted Genome Editing Using Site-Specific Nucleases*,
DOI 10.1007/978-4-431-55227-7_10

10.1 Introduction

Mice make one of the most ideal model organisms for investigating genetic features. The similarity of the mouse and human genome, the relatively short reproductive cycle and the ease of housing mouse colonies allows investigation into the roles of genes and gene products with a high homology to the human genome (Church et al. 2009; Jamsai and O'Bryan 2011). Until 2013 the main means of genetic manipulation was via homologous recombination (HR) in mouse embryonic stem (ES) cells and generation of a chimera to transmit the mutant gene. This procedure was both expensive and labour intensive, requiring highly trained researchers and technicians (Capecchi 2005). The arrival of engineered nucleases such as zinc finger nucleases (ZFNs) and transcription activator-like effector nucleases (TALENs) changed the field of genetic technology by allowing more specific targeting of genes as well as more gene loci, although they remain technically difficult to design and construct expression vectors, limiting the spread of their use (Cong et al. 2013; Li et al. 2013). The discovery and adaptation of the bacterial RNA-guided genome editing system known as CRISPR/Cas for use in mammalian cells has again changed the face of genetic technology (Cong et al. 2013; Mali et al. 2013a). Several CRISPR/Cas systems exist in bacteria, however the type II system from the *Streptococcus pyogenes* strain of bacteria has been used to target the mouse genome due to its use of a single Cas nuclease, Cas9 (Cong et al. 2013). The use of this system in mice was first applied by two independent research groups in 2013, one group mutating an EGFP tag at a transgenic locus (Shen et al. 2013) and one generating a mutant mouse line with insertions or deletions (known collectively as indels) at multiple loci as well as introducing targeted point mutations into endogenous genes (Wang et al. 2013). Since then the CRISPR/Cas system has been optimised for use in the mouse genome and its use has progressed to in vivo production of human disease models in less than half the time of traditional methods (Wang et al. 2013; Wu et al. 2013; Zhou et al. 2014). This chapter will describe the various ways the CRISPR/Cas9 system has been used in the mouse and discusses the benefits and future uses of these methods in furthering our ability to utilise mouse models for investigating mammalian genetics.

10.2 Design and Validation of CRISPR/Cas9 System In Vitro

10.2.1 Initial Design and Validation Methods

CRISPR/Cas9 was first applied to mammalian cells by Cong et al. (2013). In order to investigate whether this new system could be adapted from its bacterial origin (Garneau et al. 2010) and used in a similar manner, Cong et al. (2013) developed what is now the basic CRISPR/Cas9 delivery tool to generate genetic manipulations. This delivery tool was a vector that co-expressed the codon optimized Cas9 (hCas9) nuclease along with the designed guide RNA (gRNA) under the expression of the hybrid chicken beta actin promoter (CBh) and human U6 promoter, respectively (http://www.genome-engineering.org/crispr/; Cong et al. 2013). They confirmed the flexibility of this vector in mouse neuro2A (N2A) cell lines and named it the pX330 plasmid (http://www.addgene.org/42230/). This vector became the basic tool for CRISPR/Cas9 delivery in various mammalian organisms.

CRISPR/Cas9 delivery was further optimised and used in mice by Shen et al. (2013). This research group transcribed the Cas9 mRNA and gRNA in vitro, injected fertilised oocytes, cultured to the blastocyst stage to assess toxicity of the delivery system then extracted the DNA and performed a cleavage assay (described by Cho et al. 2013) to determine the success of CRISPR/Cas9 disruption at the target locus. This method became the basic validation method for all studies to follow. Once Shen et al. (2013) established the validation protocol they utilised the already available Pouf5-IRES-EGFP knockin mouse line, designing gRNAs targeting the allele containing the EGFP gene locus. This was the first time CRISPR/Cas9 was used in the mouse. By co-injecting their optimised Cas9 mRNA and gRNA into the cytoplasm of fertilised oocytes and transferring the surviving 2-cell embryos to pseudopregnant female mice they achieved pups heterozygous for the knockin Pouf-EGFP gene (as expected) with 1/5 pups containing a further mutation at the gRNA target site (Shen et al. 2013). This was repeated in the CAG-EGFP knockin mouse line with 1/7 pups demonstrating disruption at the target locus.

Further studies using the protocols established by Cong et al. (2013), Mali et al. (2013a) and Shen et al. (2013) saw the validation of the CRISPR gRNAs utilising human and mouse ES cells. These methods were time consuming, technically difficult and required the sacrifice of many mice. Once DNA was extracted from these cell types assessment of the efficiency of each individual gRNA came through more time consuming and expensive assays, the most common being T7EN1 and surveyor cleavage (Cong et al. 2013; Fujii et al. 2013; Ran et al. 2013; Shen et al. 2013; Sung et al. 2014; Wang et al. 2013; Yang et al. 2013).

10.2.2 Simpler and Faster Validation Method

In order to decrease the cost of the technique and in animal lives, as well as ensure the spread of this CRISPR/Cas9 system, a simpler, faster method of validating gRNA cleavage ability was necessary. Our research group developed an assay whereby the DNA cleavage ability of individual gRNAs was assessed by observing EGFP fluorescence in HEK293T cells, a cell line that is far easier to culture and maintain than ES cells whilst also circumventing the need to use blastocyst stage embryos – saving mouse lives (Mashiko et al. 2013a). This method also avoided the need for DNA sequencing, resulting in a saving on costs and time from design of the gRNA to its validation in vitro (Fig. 10.1).

The simple EGFP assay method involved PCR amplification of an approximately 300–1,000 bp sequence of the genome, with the gRNA target loci at the centre, insertion of this segment into an EGFP expression vector such that the EGFP expression would be disrupted (pCAG-EG-target-FP). Thus, the prepared reporter pCAG-EG-target-FP vector would be used to co-transfect HEK293T cells simultaneously with the

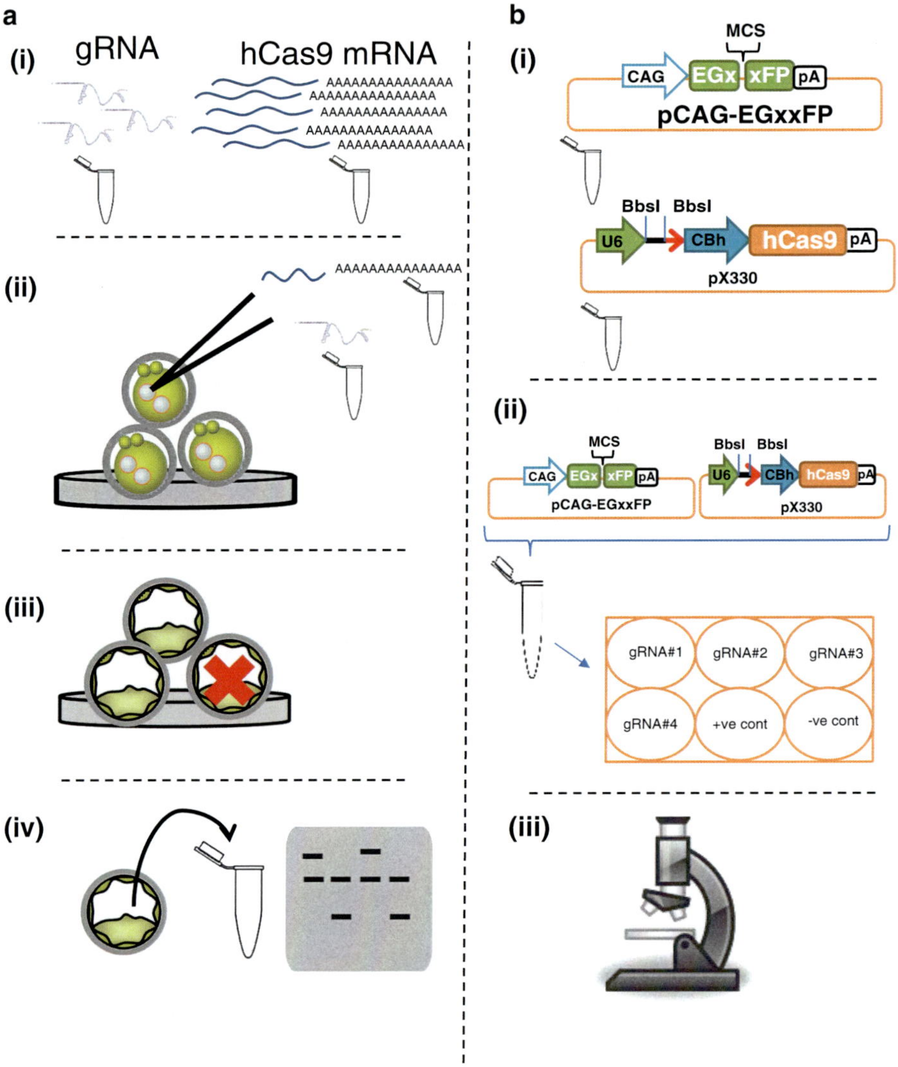

Fig. 10.1 Scheme of CRISPR/Cas9 validation in embryo or cultured cells (modified from Mashiko et al. 2013b) (**a**) (*i*) Generation of Cas9 mRNA and designed gRNA. (*ii*) Microinjection into fertilised one cell embryos. (*iii*) Culturing to blastocyst stage to assess toxicity. (*iv*) Extraction of DNA and performing cleavage assay to assess gRNA activity. (**b**) (*i*) Design and insertion of gRNA into pX330 plasmid. (*ii*) Insertion of target sequence into pCAG-EGxxFP plasmid and transfection of cultured cells. (*iii*) Analysis of EGFP fluorescence under microscope determines activity of gRNA

basic Cas9/gRNA expression plasmid pX330 carrying individual gRNA (Fig. 10.1). After 48 hours of incubation, the level of fluorescence indicates the efficiency of that particular gRNA to direct Cas9 to the target loci for cleavage (Fig. 10.2) (Mashiko et al. 2013a).

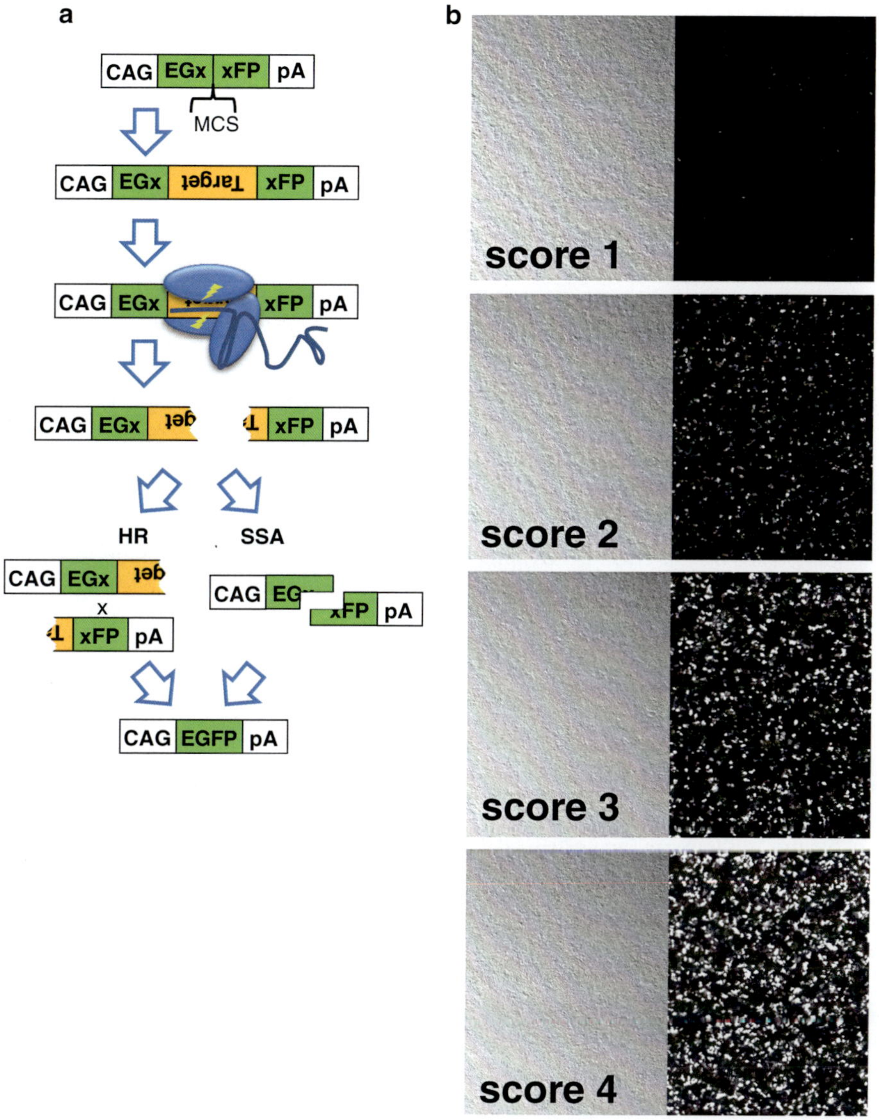

Fig. 10.2 Efficiency of gRNA as determined by EGFP fluorescence assay (modified from Mashiko et al 2013a). (**a**) Insertion of target sequence into pCAG-EGxxFP vector, transfection with pX330 containing designed gRNA, cleavage of target region, HR (Homologous Recombination) or SSA (Single Strand Annealing) repair resulting in **b**. Various levels of fluorescence: *score 1*, dark; *score 2*, darker than control; *score 3*, same as control; *score 4*, brighter than control

The gRNA producing the strongest EGFP signals were chosen for microinjection into the pronucleus of fertilised oocytes. A follow-up study demonstrated that the higher the level of fluorescence observed in the HEK293T cells, the more likely gene modified organism (GMO) pups were obtained for an individual gRNA

(Mashiko et al. 2013b). This method is not only faster and simpler to perform than the previous protocols, but it also results in a saving on mouse lives and expense, as embryos are not wasted in testing gRNA efficiency.

10.3 Oocyte Injection

Delivery of the CRISPR/Cas9 system to the mouse oocyte depends on the type of genetic disruption desired. Although Cas9 can be delivered as protein, greater efficiency in generating GMO mice is obtained when DNA or RNA delivery systems are used (Wang et al. 2013; Mashiko et al. 2013b).

10.3.1 RNA Method

The use of RNA injection is advantageous for generating subtle genetic manipulations such as point mutations or insertion of small nucleotide sequences (e.g. restriction enzyme sequences). Whilst different research groups have slightly different protocols for Cas9 mRNA and gRNA production the overall method is very similar (Cong et al. 2013; Fujii et al. 2013; Ran et al. 2013; Shen et al. 2013; Sung et al. 2014; Wang et al. 2013; Wu et al. 2013; Zhou et al. 2014).

Cas9 mRNA is produced via use of a commercial kit. The Cas9 mRNA is then modified via capping of the 5′ region and polyadenylation of the 3′ region to closer resemble biologically produced Cas9 mRNA. To ensure the Cas9 protein is taken up by the nucleus of the fertilised oocyte a nuclear localisation signal (NLS) is attached to Cas9 via a linker sequence (Mali et al. 2013a; Shen et al. 2013). Similarly, gRNA is produced via a commercial kit and both Cas9 mRNA and gRNA are stored at −80 °C until microinjection into the cytoplasm or pronuclei of the fertilised oocyte.

This method is time consuming and technically difficult. A simpler method was developed by Mashiko et al. (2013a) and is outlined in Sect. 10.3.2.

10.3.2 Plasmid Method

Mashiko et al. (2013a) found that microinjection of the pX330 plasmid coexpressing hCas9 and gRNA into the pronucleus of fertilised oocytes was an equally effective method of CRISPR delivery.

This plasmid is readily available and only requires insertion of the annealed oligonucleotides against target DNA. Validation in vitro is simple and can be achieved in a matter of days (validation method described in Sect. 10.2.2). Delivery of the plasmid requires a single microinjection to either the pronucleus or the cytoplasm of the fertilised oocyte (Mashiko et al. 2013a).

10.3.3 Discussion of RNA vs Plasmid

Whilst direct comparison between the two methods cannot be made due to differences in toxicity and stability of the molecules as well as differences in injection site (cytosol versus the pronucleus) enough mice have been generated via both methods to demonstrate they are at least equally effective (Mashiko et al. 2013a, b). Using the basic plasmid method is fast, economically advantageous, reliable and simple to perform, making it an ideal starting point for production of gene disrupted mice. The plasmid method also has the advantage of reducing integration of the Cas9 endonuclease itself into the target genome, as it is delivered as a circular vector rather than linear DNA (Mashiko et al. 2013a). At the time of writing, the injection of the plasmid has been used mainly to produce straight knockouts (whereby the entire gene is ubiquitously deleted from the mouse); with fewer cases of introduction of point mutations or small tags (data not shown). For the majority of situations requiring insertion of a point mutation or other complicated genetic manipulation the RNA method of delivery may be more effective.

The efficiency of both methods can be controlled by adjusting the concentration of plasmid/RNA injected. It has been demonstrated that increasing the concentration of plasmid/RNA increases the likelihood of obtaining GMO pups (Fujii et al. 2013; Mali et al. 2013a). For those using the RNA method it is a simple matter to increase the concentration of RNA (can be higher than 100 ng/μl), as RNA is not toxic to the developing embryo; however DNA is toxic to embryos at higher concentrations, limiting the use of the plasmid method (usually 5-10 ng/μl). The use of RNA, however, whilst technically more difficult might be advantageous when creating point mutations or inserting new genomic material into an inefficient target locus.

These minor differences aside, comparisons between the two methods (as far as they can be compared) have shown both are equally effective at delivering the CRISPR/Cas9 system to the mouse embryo. The decision as to which method to use is entirely dependent on the resources available and the type of genetic manipulation desired to be achieved.

10.4 Oligo-Mediated Genetic Modifications

10.4.1 Point Mutations

The introduction of point mutations is not a new concept. Traditional methods utilising ES cells to introduce point mutations are difficult but not impossible. These methods involve multiple rounds of HR, insertion of a negative selection marker and replacement with donor DNA carrying the desired point mutation, but never became widespread due to their complexity, expense and often-times low efficiency (Askew et al. 1993; Hasty et al. 1991; Menke 2013; Stacey et al. 1994; Valancius et al. 1991; Wu et al. 1994). Conversely, the first instance of the CRISPR/Cas9 system being used to generate a point mutation in the mouse resulted in an efficiency of

~80 % mutation (Wang et al. 2013). The introductions of point mutations or small tags are also being performed via the pX330 plasmid injection method (Mizuno et al. 2014).

By allowing the introduction of point mutations instead of whole gene deletion, researchers can now analyse the functional importance of protein domains, alternatively spliced exons, phosphorylation sites, cis-regulatory elements, enhancers and promoters, transcription factor binding sites and many other factors in vivo (Menke 2013; Wang et al. 2013).

The first example of this in mice was done by Wang et al. (2013) as they endeavoured to solve the issue of predicting what indels CRISPR/Cas9 cleavage would produce. This was achieved utilising the RNA method to change a SacI restriction enzyme site to that of EcoRI in the Tet1 gene, and EcoRV to EcoRI in the Tet2 gene. Analysis of injected embryos revealed one or both mutations had occurred and transplantation of embryos to pseudopregnant females produced pups with the desired mutation in one or both target sites (Wang et al. 2013).

To create a point mutation or insertion of a small nucleotide sequence an oligonucleotide homologous to the sequence spanning the CRISPR cleavage site and containing the desired mutation/insertion is co-injected with the Cas9 mRNA and gRNA (Fig. 10.3). This oligonucleotide shall henceforth be referred to as the "donor oligo". Yang et al. (2013) trialled three methods of delivery for the Cas9/gRNA and the donor oligo; simultaneous microinjection of all three components into the cytoplasm of fertilised oocytes, simultaneous microinjection of all three components into the pronucleus of fertilised oocytes and microinjection of Cas9 mRNA/gRNA into the cytoplasm, followed two hours later by microinjection of the donor oligo into the pronucleus of fertilised oocytes. They found that simultaneous microinjection

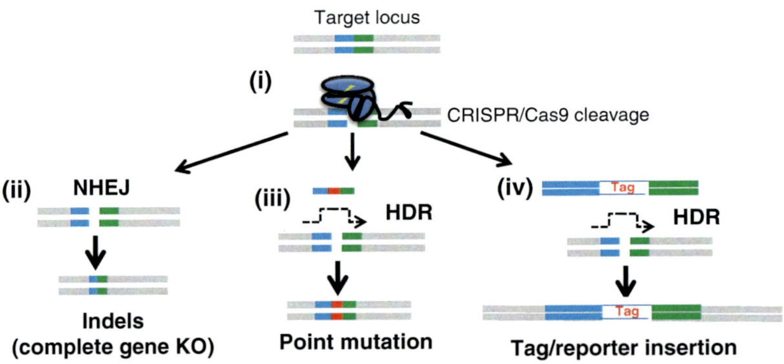

Fig. 10.3 NHEJ vs HR method of gene disruption (modified from Mashiko et al. 2013a). (**i**) Target loci with homologous region on either side (*blue* and *green*) is cleaved by the CRISPR/Cas9 system. (**ii**) Without co-injection CRISPR/Cas9 creates a DSB repaired by NHEJ resulting in indels. This generates a complete gene knockout. (**iii**) By co-injecting a donor oligo containing homologous sequences to the region surrounding the target locus (*blue* and *green*) as well as the desired mutation (*red*) HR will generate a point mutation. (**iv**) By co-injecting double stranded DNA (dsDNA) containing a larger tag or reporter sequence (*blue* and *green* indicate sequence homologous to the region surrounding the target locus)

into the cytoplasm or pronucleus were equally effective (9-19 % efficiency), but separate microinjection into the cytoplasm followed by the pronucleus had a very low efficiency (0-3 %) (Yang et al. 2013). It is most likely that this is due to the extra interference with the oocyte themselves, by disrupting several membranes over two hours rather than one or both membranes at once then allowing the embryos to recover.

10.4.2 Inserting Tags

The oligonucleotide methodology can also be used to insert small nucleotide sequences such as reporters or tags. These reporters/tags greatly improve the ease with which genetic manipulations can be observed. Yang et al. (2013) first attempted this by fusing the 42 nt V5-tag to the last codon of the *Sox2* gene. This was done utilising the oligo-mediated mutation method (discussed in Sect. 10.4.1), but instead of a donor oligonucleotide containing a point mutation it contained a larger segment of genetic sequence to insert—the V5-tag sequence. The addition of this particular tag allowed for easier analysis of the GMO mouse line—the V5 antibody is highly specific and its addition to the target gene allowed its localisation to be observed more easily (Yang et al. 2013). Embryos were allowed to develop into blastocysts which were then explanted into culture from which ES cells were derived. Sequence analysis of DNA from ES cells as well as western blot analysis confirmed integration of the V5-tag to the *Sox2* gene (Yang et al. 2013). However it should be noted that the size of the oligonucleotide to be injected is limited – commercially most companies only supply oligonucleotides of approximately 150 nt, though lengths of up to 250 nt can be sourced. This limits the size of the sequence to be inserted as there needs to be 30-50nts on either side of the insertion site homologous to the wild type target genome.

The CRISPR/Cas9 system has proven effective, efficient, rapid and economically viable for producing oligo-mediated point mutations as well as insertion of reporter/tag sequences. It far surpasses the previous methods of ES cell HR, saving on time, expenditure and mouse lives. The ability of the CRISPR/Cas9 system to target genes in the same family, gene clusters and other previously difficult areas to target solidifies it as an extremely versatile technology.

10.5 dsDNA Mediated Homologous Recombination—Generation of Knockin Mice

As well as insertion of small nucleotide sequences (discussed in Sect. 10.4) the CRISPR/Cas9 system, along with a double stranded DNA, can be used to insert larger sequences (Yang et al. 2013). The first example of this was the insertion of the transgenic marker mCherry into the C-terminal of the *Nanog* gene (Yang et al. 2013).

In this case Yang et al. (2013) utilised the plasmid method, to reduce the risk of integration of the donor dsDNA into the target locus. Fertilised oocytes were co-injected and cultured to the blastocyst stage, however success was limited. Higher concentrations of the vector produced high levels of mCherry signalling, but proved toxic to the blastocyst stage embryos. Low concentrations were less toxic but mCherry signalling was markedly reduced. Embryos injected with the mCherry sequence were transferred to pseudopregnant females, however less than 10 % of pups were positive for the transgenic signal. Of the pups that did survive, correct cellular location of mCherry was noted (Yang et al. 2013).

Another signalling technique trialled by Yang et al. (2013) was insertion of a larger fluorescent tag, in this case the published vector sequence IRES-EGFP-LoxP-Neo-LoxP expression cassette into the 3′ UTR of the *Oct4* gene. Not only did this assess the ability of CRISPR to insert a larger nucleotide sequence, it was also the first time a genetic sequence containing LoxP sites was successfully integrated both in vitro and in vivo. Success rates of pups expressing GFP at the correct target site was similar to that of V-5 tag, slightly higher than that of the mCherry transgene (Yang et al. 2013).

10.5.1 Conditional Knockouts/Knockins

Initial studies of the CRISPR/Cas9 system have developed a technology that, whilst extremely versatile, still remains limited in some areas. This includes that CRISPR/Cas9, in its basic form generates mice with ubiquitous genetic manipulations, and at the time of writing its use to generate conditional knockouts/knockins is not widespread. Traditional methods of genetic manipulation allow for conditional knockouts through use of the Cre/LoxP or Flp/Frt systems (Menke 2013; Tamowski et al. 2010). This is highly useful for analysing gene functions in certain tissues or at certain stages of the life cycle (Dymecki 1996; Feil et al. 1996; Gossen and Bujard 1992; Sauer et al. 1987; Tamowski et al. 2010). The difficulty in adapting CRISPR for this purpose lay in the need for two LoxP sites to be integrated in the same allele in the correct orientation, creating a floxed gene. This has been achieved through the design of four gRNAs for the *Mecp2* gene; two for intron 2 and two for exon 3 (Yang et al. 2013). Two donor oligos were designed covering each pair of CRISPR sites and containing LoxP sites. To make it easier to determine insertion of each donor oligo NheI and EcoRI restriction enzyme sites were added to intron 2 and exon 3 respectively. Utilising the RNA method of CRISPR/Cas9 delivery and validating gRNA efficiency in blastocyst stage embryos Yang et al. (2013) obtained pups with integration of both LoxP sites onto the same allele with approximately 16 % efficiency. To ensure the floxing was functional in vitro, Cre-mediated recombination was performed. In all cases the floxed region was successfully excised (Yang et al. 2013). However, despite the success of this method a significant number of pups were produced that had no integration of the donor but did have large deletions in the target loci due to NHEJ (Non Homologous End Joining) repair of CRISPR

DSBs (Double Strand Breaks) (Yang et al. 2013). For the creation of conditionally floxed mutant mouse lines in a single generation the success rate of this method is very encouraging. Though, as with insertion of markers and fluorescent tags, the methodology needs to undergo a fair amount of optimisation before its regular use in the laboratory.

10.6 The Future of CRISPR/Cas9 Systems in Mice

As a basic tool for generation of mutant mice as well as a method for introduction of more intricate point mutations, tags and genetic markers the CRISPR/Cas9 system is extremely useful. The rapid progress in the development and improvement of CRISPR/Cas9 methodologies and their uses in mice demonstrate this systems versatility and adaptability. However, as with all new technologies, there are always more potential uses and obstacles to overcome.

10.6.1 Off-Target Mutations

The main issue with this system is the incidence of off-target cleavage. It is an important issue to take into consideration when analysing a phenotype of a mutant mouse line generated by CRISPR/Cas9 cleavage. As more mice are generated and the methodologies are refined the incidence of off-target cleavage has been reduced. Software such as Bowtie (http://bowtie-bio.sourceforge.net/index.shtml) can predict the incidence of off-target binding of individual gRNAs before they are generated, allowing researchers to select the sequences with the least off-target recognition sites in the genome (Wang et al. 2013; Mashiko et al. 2013a, b). Off-target frequency was found to be directly correlated with concentration of Cas9 mRNA being injected (Fujii et al. 2013; Mali et al. 2013a), therefore by lowering the concentration of Cas9 mRNA researchers can further reduce incidence of off-target cleavage. When generating mouse models this may be enough to minimise the effect, as analysis of off-target cleavage is relatively simple and appropriate selection of gRNAs and Cas9 concentration can reduce the number of analyses needed and any remaining potential off-target cleavages can often be segregated by breeding. However this becomes a larger issue when applying this system to humans, as any gene therapy in human tissues needs to have as minimal risk of off-target mutations as possible (Fu et al. 2013).

Another potential solution to the off-target problem is the modification of the Cas9 endonuclease. By modifying the two domains of the Cas9 protein responsible for inducing double strand breaks (DSBs) to only induce a single strand nick in the target DNA, researchers have found a way to alleviate the rate of off-target mutations (Cho et al. 2014; Ran et al. 2013; Mali et al. 2013b). It should be noted, however, that modified Cas9-nickase (Cas9n) requires the use of two gRNAs, each targeting a single strand of the target locus. The cutting of both strands at the target site results in a DSB, which is repaired by NHEJ or HR (resulting in the characteristic mutations induced by CRISPR/Cas9), however any off-target binding from each individual gRNA results in a single stranded nick, which is easily repaired with high fidelity by base excision repair mechanisms (Cho et al. 2014; Mali et al. 2013b; Ran et al. 2013). Indeed the use of modified Cas9n has been demonstrated to prevent chromosomal rearrangements such as deletions, translocations and inversions (Cho et al. 2014). Whilst this technique is beginning to take hold it still remains to be developed for widespread use in the generation of mouse models of disease. As off-target mutation occurs in approximately 1/100 mutant mice this is not an area of great concern (Mashiko et al. 2013a). In adapting the CRISPR/Cas9 system for use in human cells and gene therapy the off-target frequency becomes of vast importance.

10.6.2 Other Technologies

With its increased simplicity, scientists are now combining the CRISPR/Cas9 system with novel and existing genetic technologies to further investigate novel aspects of the mouse genome.

10.6.2.1 CRISPR Mediated Gene Activation/Inactivation

The CRISPR/Cas9 system utilises Watson-Crick base pairing to identify and bind to the target sequence. It can therefore target practically any region of the genome (Chen et al. 2013; Qi et al. 2013). Initial research into the ability of inactivated Cas9 (achieved by mutating the two catalytic domains and named dCas9) found that attachment of the CRISPR/dCas9 complex to the genome of bacteria blocked binding of RNA polymerase, halting gene transcription (Qi et al. 2013). This was applied to mammalian cells with moderate success, however design and selection of targeting gRNAs has more of an effect on binding to the target locus in mammalian cells than in bacteria (Qi et al. 2013). This research was built upon further by Gilbert et al. (2013) when they investigated fusion of an effector protein to the dCas9, resulting in targeted repression/activation of specific loci. The fused proteins were found to mimic biological reactions and the targeting ability of dCas9 was as efficient as the active form. Mali et al. (2013a) then investigated the possibility of fusing a regulatory sequence to the dCas9. By fusing the transcriptional regulator to the C terminus of dCas9 and co-transfecting with a gRNA targeting the promoter sequence for a particular gene, up-regulation of a target gene in human cells is possible. However, whilst fusing the effector protein directly to the gRNA, does show some success, this requires the simultaneous addition of a separate fusion protein. Overall fusion of the effector protein to dCas9 works 1.5-3 times more efficiently and requires less handling than fusion to the gRNA (Mali et al. 2013a).

This technology, whilst extremely promising is yet to be tested in living organisms. As such, mice would be an ideal candidate organism for investigating this type of CRISPR interference (CRISPRi) in vivo.

10.6.2.2 Observing the Genome

Visualisation of individual genes in live cells has proven difficult with the technology currently available (Chen et al. 2013). As function is directly linked to spatial organisation it is imperative that the localisation of certain genetic elements is understood. Current technology can visualise some genetic elements, such as repetitive sequences, telomeres and centromeres (Chen et al. 2013; Hellwig et al. 2008; Wang et al. 2008). Despite this, imaging methods such as fluorescence in situ hybridisation (FISH) cannot be used in living cells due to fixation techniques and DNA denaturation (Chen et al. 2013). A similar type of CRISPRi to that discussed in Sect. 10.6.2.1 is the use of dCas9 to deliver a tag or reporter molecule directly to a specific gene locus. This use of CRISPRi was first attempted by Chen et al. (2013) by utilising EGFP-fused to dCas9 and structurally optimised gRNAs targeting the MUC4 loci in human cells, with moderate success. As this point in time a basic protocol has been developed to visualise native chromatin organisation and dynamics in living human cells—a non-destructive methodology allowing observation of changes over a long period of time (Chen et al. 2013). Whilst requiring further investigation and development the use of tagged-dCas9 opens up avenues to observe cellular mitosis, including sister chromatid cohesion, chromosome condensation/decondensation as well as homologous pairing and recombination and the spatial organisation and dynamic interactions of chromatin and other molecules in real time in living cells (Chen et al. 2013). As with the epigenetic modification (discussed in Sect. 10.6.2.1) these methods have yet to be trialled in a living organism, and the mouse would be an ideal candidate for initial investigations of this technique in vivo.

10.6.2.3 Gene Modification to Correct Impairment

CRISPR has also been used to explore the possibility of correcting genetic diseases resulting from known mutations. Wu et al. (2013) attempted to correct a dominant genetic disorder which causes cataracts—a 1 bp deletion in the *Crygc* gene. They used targeted design of gRNAs, 3 across the mutation site, 1 downstream (as the mutation causes a neo-PAM sequence) and 1 in the WT upstream untranslated region. They found through testing on ES cells that the gRNA targeting downstream and one of gRNAs targeting the region across the mutation locus caused cutting of only the mutated allele, not cutting the WT ‘healthy’ allele. With the addition of an oligonucleotide containing the ‘healthy’ genetic sequence Wu et al. (2013) found that of the two gRNAs targeting the mutated allele, the downstream gRNA promoted HR more often than the one targeted across the mutated region, thus this downstream gRNA was chosen for in vivo analysis.

Using the RNA method of delivery into the cytoplasm of fertilised oocytes and transfer of zygotes to a pseudopregnant female, approximately 45 % of pups was GMO, with 40 % of these having undergone HR (the remaining GMO's were due to NHEJ). All pups with HR had no cataracts, due to HR using the healthy sister chromatid as a template (Wu et al. 2013). The process was repeated but with the addition of a 'healthy' oligonucleotide and sequence analysis showed the mutated allele was repaired from integration of the oligo with a higher frequency than CRISPR alone; 48 % pups were GMO and 64 % of GMO pups had no cataracts. Wu et al. (2013) followed the 'healed' mice into the second generation and found that, as with the other CRISPR indels thus far produced, the correction was passed onto all offspring. Off-target analysis showed that only 2 mice out of 12 had off-target indels.

A second study by Yin et al. (2014) also used CRISPR/Cas9 to correct a genetic disease – this time hereditary tyrosinemia type I (HTI). As with the study by Wu et al. (2013) this disease is characterised by a single deletion, in this case in exon 8 of the *Fah* gene, resulting in a misfolded protein, build-up of toxic metabolites and eventually liver damage. Mice have the same mutation and disease phenotype as in humans, making them an excellent model to trial CRISPR/Cas9 genetic repair. Different to the study by Wu et al. (2013) in this case the CRISPR/Cas9 repair was attempted by delivery to adult mice – by hydrodynamic tail vein injection of the plasmid method of CRISPR delivery coupled with an oligonucleotide containing the 'healthy' WT sequence (Yin et al. 2014). Mice injected with the donor oligo, pX330 and one of three designed gRNAs were successfully treated to varying degrees (each gRNA having different efficiency), whilst mice injected with saline, donor oligo, pX330 but no gRNA all developed HTI and had to be euthanized (Yin et al. 2014). This disease has a strong positive selection for healthy/healed cells, whereby healthy cells will replicate and eventually replace the mutated/damaged cells of the liver. Yin et al. (2014) concluded that genetic diseases that display this type of positive selection for healthy/healed cells are particularly suitable for this type of gene-repair, however they do note that their method of delivery (tail injection) does not translate well to the clinical setting, and much more work is still needed.

Conclusion

The ability of the CRISPR/Cas9 system to be used to generate delicate genetic manipulations in specific regions of the mouse genome has opened a whole new window into investigating genomics. The CRISPR/Cas9 system is so versatile that a single set of designed gRNAs can be used with various forms of the Cas9 protein – nuclease Cas9 for genomic engineering, dCas9 fused with effector proteins for gene regulation control or dCas9 fused with tags/reporter proteins for visualisation and analysis of genetic elements in live cells (Chen et al. 2013). The mouse genome with its similarity to humans remains a valuable organism for the study of human disease conditions and with the advancement of the CRISPR/Cas system of gene manipulation can now be used to investigate finer details of gene mechanisms, localisation and function.

References

Askew GR, Doetschman T, Lingrel JB (1993) Site-directed point mutations in embryonic stem cells: a gene-targeting tag-and-exchange strategy. Mol Cell Biol 13:4115–4124

Capecchi MR (2005) Gene targeting in mice: functional analysis of the mammalian genome for the twenty-first century. Nat Rev Genet 6:507–512

Chen B, Gilbert LA, Cimini BA, Schnitzbauer J, Zhang W, Li GW, Park J, Blackburn EH, Weissman JS, Qi LS, Huang B (2013) Dynamic imaging of genomic loci in living human cells by an optimized CRISPR/Cas system. Cell 155:1479–1491

Cho SW, Kim S, Kim JM, Kim JS (2013) Targeted genome engineering in human cells with the Cas9 RNA-guided endonuclease. Nat Biotechnol 31:230–232

Cho SW, Kim S, Kim Y, Kweon J, Kim HS, Bae S, Kim JS (2014) Analysis of off-target effects of CRISPR/Cas-derived RNA-guided endonucleases and nickases. Genome Res 24:132–141

Church DM, Goodstadt L, Hillier LW, Zody MC, Goldstein S, She X, Bult CJ, Agarwala R, Cherry JL, DiCuccio M et al (2009) Lineage-specific biology revealed by a finished genome assembly of the mouse. PLoS Biol 7:e1000112

Cong L, Ran FA, Cox D, Lin S, Barretto R, Habib N, Hsu PD, Wu X, Jiang W, Marraffini LA, Zhang F (2013) Multiplex genome engineering using CRISPR/Cas systems. Science 339:819–823

Dymecki SM (1996) Flp recombinase promotes site-specific DNA recombination in embryonic stem cells and transgenic mice. Proc Natl Acad Sci U S A 93:6191–6196

Feil R, Brocard J, Mascrez B, Lemeur M, Metzger D, Chambon P (1996) Ligand-activated site-specific recombination in mice. Proc Natl Acad Sci U S A 93:10887–10890

Fu Y, Foden JA, Khayter C, Maeder ML, Reyon D, Joung JK, Sander JD (2013) High-frequency off-target mutagenesis induced by CRISPR-Cas nucleases in human cells. Nat Biotechnol 31:822–826

Fujii W, Kawasaki K, Sugiura K, Naito K (2013) Efficient generation of large-scale genome-modified mice using gRNA and CAS9 endonuclease. Nucleic Acids Res 41:e187

Garneau JE, Dupuis ME, Villion M, Romero DA, Barrangou R, Boyaval P, Fremaux C, Horvath P, Magadán AH, Moineau S (2010) The CRISPR/cas bacterial immune system cleaves bacteriophage and plasmid DNA. Nature 468:67–71

Gilbert LA, Larson MH, Morsut L, Liu Z, Brar GA, Torres SE, Stern-Ginossar N, Brandman O, Whitehead EH, Doudna JA, Lim WA, Weissman JS, Qi LS (2013) XCRISPR-mediated modular RNA-guided regulation of transcription in eukaryotes. Cell 154:X442–X451

Gossen M, Bujard H (1992) Tight control of gene expression in mammalian cells by tetracycline-responsive promoters. Proc Natl Acad Sci U S A 89:5547–5551

Hasty P, Ramirez-Solis R, Krumlauf R, Bradley A (1991) Introduction of a subtle mutation into the Hox-2.6 locus in embryonic stem cells. Nature 350:243–246

Hellwig D, Münch S, Orthaus S, Hoischen C, Hemmerich P, Diekmann S (2008) Live cell imaging reveals sustained centromere binding of CENP-T via CENP-A and CENP-B. J Biophotonics 1:245–254

Jamsai D, O'Bryan MK (2011) Mouse models in male fertility research. Asian J Androl 13:139–151

Li W, Teng F, Li T, Zhou Q (2013) Simultaneous generation and germline transmission of multiple gene mutations in rat using CRISPR-Cas systems. Nat Biotechnol 31:684–686

Mali P, Yang L, Esvelt KM, Aach J, Guell M, DiCarlo JE, Norville JE, Church GM (2013a) RNA-guided human genome engineering via Cas9. Science 339:823–826

Mali P, Aach J, Stranges PB, Esvelt KM, Moosburner M, Kosuri S, Yang L, Church GM (2013b) CAS9 transcriptional activators for target specificity screening and paired nickases for cooperative genome engineering. Nat Biotechnol 31:833–838

Mashiko D, Young SA, Muto M, Kato H, Nozawa K, Ogawa M, Noda T, Kim YJ, Satouh Y, Fujihara Y, Ikawa M (2013a) Feasibility for a large scale mouse mutagenesis by injecting CRISPR/Cas plasmid into zygotes. Dev Growth Differ 26:12113

Mashiko D, Fujihara Y, Satouh Y, Miyata H, Isotani A, Ikawa M (2013b) Generation of mutant mice by pronuclear injection of circular plasmid expressing Cas9 and single guided RNA. Sci Rep 3:3355

Menke DB (2013) Engineering subtle targeted mutations into the mouse genome. Genesis 51:605–618

Mizuno S, Dinh TT, Kato K, Mizuno-Iijima S, Tanimoto Y, Daitoku Y, Hoshino Y, Ikawa M, Takahashi S, Sugiyama F, Yagami K (2014) Simple generation of albino C57BL/6J mice with G291T mutation in the tyrosinase gene by the CRISPR/Cas9 system. Mamm Genome 25:327–334

Qi LS, Larson MH, Gilbert LA, Doudna JA, Weissman JS, Arkin AP, Lim WA (2013) Repurposing CRISPR as an RNA-guided platform for sequence-specific control of gene expression. Cell 152:1173–1183

Ran FA, Hsu PD, Lin CY, Gootenberg JS, Konermann S, Trevino AE, Scott DA, Inoue A, Matoba S, Zhang Y, Zhang F (2013) Double nicking by RNA-guided CRISPR Cas9 for enhanced genome editing specificity. Cell 154:1380–1389

Sauer B (1987) Functional expression of the cre-lox site-specific recombination system in the yeast *Saccharomyces cerevisiae*. Mol Cell Biol 7:2087–2096

Shen B, Zhang J, Wu H, Wang J, Ma K, Li Z, Zhang X, Zhang P, Huang X (2013) Generation of gene-modified mice via Cas9/RNA-mediated gene targeting. Cell Res 23:720–723

Stacey A, Schnieke A, McWhir J, Cooper J, Colman A, Melton DW (1994) Use of double-replacement gene targeting to replace the murine alpha-lactalbumin gene with its human counterpart in embryonic stem cells and mice. Mol Cell Biol 14:1009–1016

Sung YH, Kim JM, Kim HT, Lee J, Jeon J, Jin Y, Choi JH, Ban YH, Ha SJ, Kim CH, Lee HW, Kim JS (2014) Highly efficient gene knockout in mice and zebrafish with RNA-guided endonucleases. Genome Res 24:125–131

Tamowski S, Aston KI, Carrell DT (2010) The use of transgenic mouse models in the study of male infertility. Syst Biol Reprod Med 56:260–273

Valancius V, Smithies O (1991) Testing an "in-out" targeting procedure for making subtle genomic modifications in mouse embryonic stem cells. Mol Cell Biol 11:1402–1408

Wang X, Kam Z, Carlton PM, Xu L, Sedat JW, Blackburn EH (2008) Rapid telomere motions in live human cells analyzed by highly time-resolved microscopy. Epigenetics Chromatin 1:1756–8935

Wang H, Yang H, Shivalila CS, Dawlaty MM, Cheng AW, Zhang F, Jaenisch R (2013) One-step generation of mice carrying mutations in multiple genes by CRISPR/Cas-mediated genome engineering. Cell 153:910–918

Wu H, Liu X, Jaenisch R (1994) Double replacement: strategy for efficient introduction of subtle mutations into the murine Col1a-1 gene by homologous recombination in embryonic stem cells. Proc Natl Acad Sci U S A 91:2819–2823

Wu Y, Liang D, Wang Y, Bai M, Tang W, Bao S, Yan Z, Li D, Li J (2013) Correction of a genetic disease in mouse via Use of CRISPR-Cas9. Cell Stem Cell 13:659–662

Yang H, Wang H, Shivalila C, Cheng A, Shi L, Jaenisch R (2013) One-step generation of mice carrying reporter and conditional alleles by CRISPR/Cas-mediated genome engineering. Cell 153:1370–1379

Yin H, Xue W, Chen S, Bogorad RL, Benedetti E, Grompe M, Koteliansky V, Sharp PA, Jacks T, Anderson DG (2014) Genome editing with Cas9 in adult mice corrects a disease mutation and phenotype. Nat Biotechnol 32:551–553

Zhou J, Shen B, Zhang W, Wang J, Yang J, Chen L, Zhang N, Zhu K, Xu J, Hu B, Leng Q, Huang X (2014) One-step generation of different immunodeficient mice with multiple gene modifications by CRISPR/Cas9 mediated genome engineering. Int J Biochem Cell Biol 46:49–55

Chapter 11
Genome Editing in Mice Using TALENs

Tomomi Aida

Abstract Genetically modified animals such as knockout mice are essential for elucidating in vivo gene functions and identifying genetic contributions to the molecular pathophysiology of human diseases. For the past two decades, knockout mice have been created via embryonic stem (ES) cell-based gene targeting, a time-consuming, laborious, inefficient, and expensive process. The rapid emergence of targeted genome editing technologies is drastically revolutionizing this situation. Genome editing mediated by transcription activator-like effector (TALE) nucleases (TALENs), one of the popular genome editing tools, is a simple and powerful gene-targeting technology. With its extremely high efficiency, the mouse genome can be manipulated directly in fertilized eggs without any targeting vector or selection steps by a process called in vivo genome editing. TALEN-mediated in vivo genome editing provides an exciting opportunity for simple, convenient, and ultra-rapid production of precisely targeted knockout and knockin mice. Using this technology, researchers can freely and routinely manipulate mouse genomes and accelerate in vivo functional genomic research.

Keywords Glaucoma • Glutamate transporter • In vivo • Knockin • Knockout • Mouse • TALEN

11.1 Introduction

Historically, the development of genetic manipulation technologies has opened a new era for life sciences and drastically improved our understanding of, in particular, the in vivo gene functions of living organisms. Overexpression and knock-down of genes with transgenic and RNA interference (RNAi) technologies permits the control of specific gene dosage in vivo in many species from bacteria to animals to plants. However, these technologies are relatively coarse-grained with respect to

T. Aida (✉)
Laboratory of Molecular Neuroscience, Medical Research Institute,
Tokyo Medical and Dental University, Tokyo, Japan
e-mail: aida.aud@mri.tmd.ac.jp

T. Yamamoto (ed.), *Targeted Genome Editing Using Site-Specific Nucleases*,
DOI 10.1007/978-4-431-55227-7_11

control of gene expression. In transgenic animals, the expression state of a transgene differs from that under physiological conditions. The expression level of a transgene is much higher than that in the wild type, and it is constitutively active. The expression pattern is controlled by a tissue- or cell-specific or general promoter, but the expression pattern is strongly affected by the genetic locus where the transgene is inserted; this phenomenon is known as the positional effect. Thus, transgenic animals are useful tools, but often unrepresentative of physiological and pathophysiological conditions (Saito et al. 2014). RNAi-mediated knock-down technologies can be applied to living organisms by direct delivery of small interfering RNA (siRNA), a plasmid expressing short hairpin RNA (shRNA), or as a transgenic animal expressing shRNA. RNAi provides a method for specifically downregulating the expression of a target gene. The major problems of this technology are off-target effects and incompleteness of protein depletion (Echeverri et al. 2006).

In contrast to these transgenic and RNAi technologies, ES cell-based gene targeting allows specific and precise manipulation of the genome. The mouse is the most commonly used animal model in basic research because of the unexampled availability of homologous recombination in ES cells, an essential and core step in gene-targeting technology for precise replacement of a target gene with a selectable marker that can contribute to the organism and germ cells. Owing to the first success of gene targeting in the mouse, thousands of knockout and knockin mice have been created, revealing the in vivo functions of genes. Moving ahead, The International knockout Mouse Consortium (IKMC) has comprehensively constructed knockout mice for all protein-coding genes (Sung et al. 2012; Menke 2013). IKMC has produced targeted ES cell lines, including conventional knockout, conditional knockout, and gene-trapped alleles, for more than 18,000 mouse genes in addition to mice targeted for over 2,600 loci (Skarnes et al. 2011). These genome wide and large-scale knockout mouse resources are now publicly available and researchers can obtain chimeric mice by injection of targeted ES cells of interest derived from IKMC. This allows the researchers to focus their efforts on the functional analysis of target genes, rather than on the construction of targeted ES cells and mouse lines.

Recent advances in genome-wide association studies (GWAS) and high-throughput sequencing technologies have revealed the landscape of human genetic diversity, which is comprised of tens of millions of common and rare variants, known as single nucleotide variants (SNVs), associated with health and disease (1000 Genomes Project Consortium et al. 2012; Raychaudhuri 2011). Of these, loss-of-functional SNVs including that of nonsense, splice site, or frameshift, are easy to interpret those functional consequences because patients carrying loss-of-functional SNVs are comparable to knockout mice (Veltman and Brunner 2012). In contrast, a large proportion of SNVs are misssense, silent, intronic, or intergenic SNVs, thus, functional consequences of these SNVs are hard to estimate. Thus, further efforts are required to investigate the biological consequences of these SNVs. A powerful approach to address the biological function of these SNVs is the knockin mouse carrying precisely modified human SNVs (Veltman and Brunner 2012). The knockin mouse can be used to test whether these SNVs are causal for the human phenotype of interest, thus providing unique and direct opportunities for the investigation of the functional consequence of SNVs in vivo to the complex human traits in health and disease. Südhof and colleagues showed a good example of this approach. They generated knockin mouse carrying a neuroligin-3 R451C SNV found in a subset of patients with autism, a major neurodevelopmental disorder. The neuroligin-3 R451C knockin mice showed abnormal behaviors that resembled those of human patients and abnormal synaptic transmission (Tabuchi et al. 2007). Importantly, these behavioral and electrophysiological abnormalities were not observed in neuroligin-3 knockout mouse, suggesting that the R451C SNV represents a gain-of-function mutation. Thus, knockin mouse provides a powerful approach for functional interpretation of SNVs. Although the demand for precisely modified knockin mouse is growing, Menke (2013) showed that just only 600 such mice could be found in the mutant mouse database. This situation is partially due to the several difficulties in generating such mice by conventional gene-targeting technology (Capecchi 2005).

In the conventional gene-targeting method in ES cells, generation of homozygous knockout mice generally takes more than a year and involves many processes: (1) construction of a targeting vector with long homology arms to replace a target gene with a drug selection marker, (2) electroporation of the targeting vector into ES cells and screening to isolate correctly targeted cells, (3) transplantation of correctly targeted ES cells into blastocysts, (4) obtaining and crossing F_0 chimeric mice with wild-type mice to identify germline-competent mutant lines, (5) crossing heterozygous mutants with each other, and (6) obtaining a homozygous mutant (Aida et al. 2014). Further, because ES cell lines derived from the 129/Sv mouse strain are generally used in conventional gene targeting, mutant mice must be crossed with the commonly used C57BL/6 strain mice many times to analyze the mutant in a C57BL/6 genetic background; these steps known as backcrossing generally take several years. Overall, even today, gene targeting in mouse is time consuming, laborious, inefficient, and expensive.

11.2 In Vivo Genome Editing in Mice

The recent emergence and drastic advances in targeted genome-editing technologies have opened a new era in biology, medicine, agriculture, and biotechnology (Carroll 2014; Hsu et al. 2014). Genome editing technologies enable the modification of almost any desired sequence from any living cell or organism. The technologies are based on molecular tools, including zinc-finger nucleases (ZFNs), transcription activator-like effector (TALE) nucleases (TALENs), and clustered, regularly interspaced, short palindromic repeats (CRISPR) and CRISPR-associated endonuclease 9 (Cas9), known as the CRISPR/Cas9 system. The principle of genome editing is generation of a DNA double-strand break (DSB) at a specific target locus by introducing genome editing tools into cells (Urnov et al. 2010; Joung and Sander 2013). DSB is subsequently repaired by two major cellular endogenous DNA damage repair pathways. These pathways are the error-prone, nonhomologous end-joining

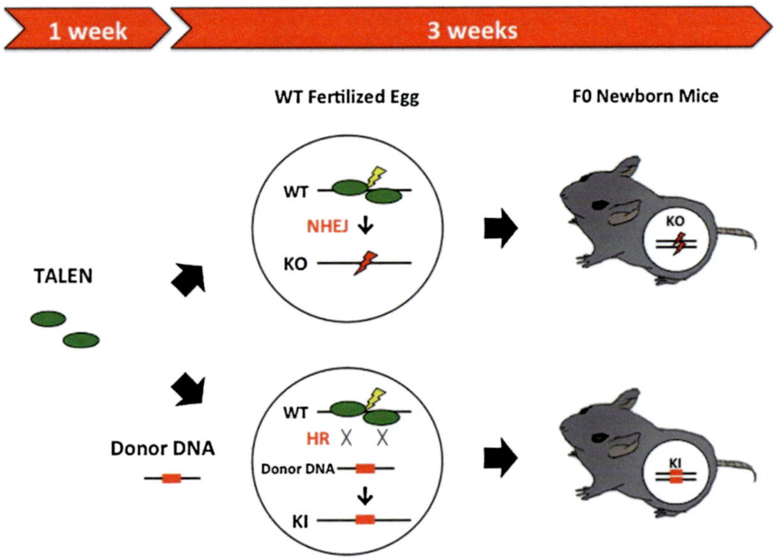

Fig. 11.1 Schematic diagram of TALEN-mediated in vivo genome editing in mouse. TALENs (*green*, 1 week for preparation) are microinjected into one-cell fertilized eggs (*circle in the middle*) derived from wildtype mice for knockout mouse production (*upper*). Donor DNA (ssOligo or targeting vector) is co-injected for knockin mouse production (*lower*). Then, DSB (*yellow knurl*) and subsequent NHEJ (*red knurl*) and/or HDR (*red box*) are induced within one-cell fertilized eggs, three weeks later, resulting in biallelically-targeted homozygous knockout (*upper*) and knockin (*lower*) mice at F_0

(NHEJ) route, which results in small deletions or sequence insertions into the DSB site, and the homology-directed repair (HDR) pathway, which relies on a donor DNA template with homology to the DSB site to achieve precise homologous recombination (Fig. 11.1). NHEJ occurs rapidly and preferentially, often leading to frameshift mutations and loss of function of the targeted genes, resulting in a gene knockout when a protein-coding sequence is targeted. HDR occurs infrequently and also results in precise and specific genome modifications such as SNV substitutions, insertions, deletions, or gene insertion, when a targeting vector or synthetic single-strand oligonucleotide (ssOligo) is co-delivered, which in turn results in a targeted knockin.

The selection steps for correctly targeted ES cells are required in conventional gene targeting in the mouse, because the efficiency of spontaneous homologous recombination is too low to perform the procedure directly in fertilized eggs (Brinster et al. 1989). The remarkable advantage of genome editing is its extremely high efficiency; NHEJ occurs in up to 50 % (or more) of cells in several primary and immortalized cells without drug selection, and HDR is stimulated by DSB and targeting efficiencies are enhanced at least 1,000-fold relative to the conventional method in several cells (Urnov et al. 2010; Hauschild-Quintern et al. 2013). Thus, one can easily imagine that by taking advantage of the highly efficient genome editing technology, researchers can directly manipulate the mouse genome in fertilized eggs (perform in vivo genome editing) without any selection or use of ES cells. The principle of in vivo genome editing involves targeting the genome by direct microinjection of the mRNA that codes for genome-editing tools into the cytoplasm or pronuclei of one-cell mouse embryos for gene knockout, and co-injection of ssOligos or targeting vectors for gene knockin mice (Fig. 11.1). In vivo genome editing is a groundbreaking technology for mutant mouse production, enabling simple and daily experiments (Menke 2013; Aida et al. 2014). Here we review gene targeting in the mouse with a focus on TALEN-mediated in vivo genome editing.

11.3 Knockout Mice

The first success of gene knockout by in vivo genome editing in mammals was achieved in the rat with ZFNs (Geurts et al. 2009), and the technique was then rapidly applied to the mouse (Carbery et al. 2010). At the same time, the simple modular DNA recognition code of TALEN was cracked, and the technology was then rapidly expanded and emerged at center stage of genome editing (Anonymous 2012). The first knockout mice created by TALEN-mediated in vivo genome editing were reported in early 2013 (Sung et al. 2013). In this impressive work, two endogenous mouse genes were targeted with extremely high efficiencies (more than 50 %) in neonatal mice. When they used high doses of TALENs, a majority of neonatal mice were biallelically targeted homozygous knockout mice. They found no off-target effects and the targeted alleles were successfully transmitted to succeeding generations. This pioneering work showed the feasibility of highly efficient TALEN-mediated in vivo genome editing in mice. Because of the ease of target-specific TALEN construction, homozygous knockout mice can be obtained within one month.

Since the first report, numerous knockout mice generated by TALEN-mediated in vivo genome editing have been reported (Aida et al. 2014). Davies and colleagues targeted the *Zic2* gene with TALENs in three different mouse strains: CD1, C3H, and C57BL/6J (Davies et al. 2013). The targeting efficiencies producing live newborns or blastocysts varied, with 10 %, 23 %, and 46 % for C57BL/6J, C3H, and CD1, respectively.

Li and colleagues generated a series of knockout mice for 10 genes, revealing the utility, convenience, and robustness of TALEN-mediated in vivo genome editing (Qiu et al. 2013). The targeting efficiencies varied from 13 to 67 %, with an average of 40 %, of live newborns. In contrast to the report of Davies et al. (2013), using one TALEN for the *Lepr* gene, which encodes the leptin receptor, they showed that the targeting efficiency did not differ between two different mouse strains (C57BL/6N and FVB/N). Among the F_0 founders, one had biallelic modifications with different frame-shift deletions and exhibited an obese phenotype resembling the Lepr mutant db/db mice phenotype. Moreover, there were no off-target effects, even at sites with only one mismatch to each TALEN. All tested F_0 founders transmitted the mutant alleles to F_1 mice with high efficiency. These results indicate that genome editing is highly accurate and efficient.

Han and colleagues generated knockout mice for the *Mlkl* gene (which encodes the mixed lineage kinase domain-like protein, essential for tumor necrosis factor-induced necrosis) by TALEN-mediated in vivo genome editing (Wu et al. 2013). They obtained 71 mutants from 390 newborns (18 % efficiency). Of these, four were homozygous mutants.

Zhou and colleagues generated knockout mice for the *Ttc36* gene (renal proximal tubule protein) by TALEN-mediated in vivo genome editing (Liu et al. 2014b). They compared the mutation efficiency between pronuclear and cytoplasmic injection and found that cytoplasmic injection works much (more than twofold) better than pronuclear injection; of these mice, a few were homozygous mutants.

Takada and colleagues applied in vivo genome editing to microRNAs (miRNAs) that are often present in intergenic regions where regulatory sequences are located (Takada et al. 2013). In this context, the advantage of in vivo genome editing is that it can delete only a target sequence, in contrast to traditional gene targeting, which leaves LoxP sequences in intergenic or intronic regions, potentially affecting transcriptional regulation. They targeted three different miRNAs (*mmu-mir-146a*, *mmu-mir-10a*, and *mmu-mir-10b*) and an intergenic region on chromosome 11, and successfully obtained each mutant mouse.

Genome-editing technologies enable gene targeting in mice, which was previously considered impossible. First, these methods permit gene targeting at loci where conventional homologous recombination cannot be applied, such as the Y chromosome. Because the Y chromosome has a unique structure containing many palindromes, conventional gene targeting in ES cells has failed. Jaenisch and colleagues targeted the *Sry* and *Uty* genes on the Y chromosome in mouse ES cells using TALENs and successfully obtained knockout mice lacking *Sry* or *Uty* (Wang et al. 2013). Takada and colleagues directly targeted the same *Sry* gene by TALEN-mediated in vivo genome editing and obtained an *Sry* knockout mouse (Kato et al. 2013). Thus, the high sequence specificity of TALENs provides a new approach for the genetic manipulation of the Y chromosome. Because *Sry* knockout male mice derived from either targeted ES cells or directly targeted fertilized eggs showed similar phenotypes such as anatomical femaleness, in vivo genome editing offers a great advantage in terms of time.

Second, in vivo genome editing allows multiple targeting of genes. Sherr and colleagues simultaneously targeted two and three genes including *Agouti*, *miR-205*, and *Arf* tumor suppressor locus by in vivo genome editing with a mixture of TALEN mRNAs against these targets. Both double and triple targeting worked well and newborns carrying mutations in two or three genes were obtained (Li et al. 2014). Previously, when researchers focused on two or more functionally redundant genes, they needed to cross each single-mutant mouse, a time-consuming and laborious process (Matsugami et al. 2006; Aida et al. 2012). Furthermore, when two genes are located next to each other on the same chromosome, double knockout mice can mostly not be obtained by crossing two single-knockout mice. Thus, researchers have generated double-targeted ES cells by sequential targeting, which is much more time consuming, laborious, and expensive than single-gene targeting (Kitajima et al. 2000). In this context, in vivo genome editing offers a great advantage for simultaneous targeting of multiple genes and provides invaluable tools for studying cooperative roles of functionally redundant or clustered genes.

Third, these methods enable gene targeting in diverse genetic backgrounds of mouse strains. In conventional gene targeting, ES cells derived from 129 mouse strains are most often used owing to the high efficiency of gene targeting. However, subsequent analyses of targeted mice in a C57BL/6 genetic background are preferred. Thus, time-consuming backcrossing, which takes at least one year, is essential. As demonstrated previously (Davies et al. 2013; Qiu et al. 2013), in vivo genome editing can be applied to any mouse strain and provides the opportunity of analyzing the targeted mice immediately without backcrossing.

11.4 Knockin Mice

Although the success of in vivo genome editing has enabled the rapid generation of knockout mice, the development of this technique for producing knockin mouse models would fully exploit its potential. As with knockout mice, earlier successes in knockin mouse production by in vivo genome editing were achieved with ZFNs and conventional targeting vectors (Meyer et al. 2010; Cui et al. 2011). The construction of traditional targeting vector for the targeted integration of SNVs by HDR is disproportionately laborious and time consuming because only one or a few nucleotides substitutions are required. The use of ssOligos as donors for HDR allows bypassing this process, because ssOligos can be chemically synthesized and purchased from many companies within a few days. Davis and colleagues systematically investigated ssOligo-mediated targeted integration of point mutations by ZFN in several human cell lines. They found that ssOligo works very well as donor template for HDR, and surprisingly, its efficiencies were up to twice those achieved using conventional targeting vectors (Chen et al. 2011). Kühn and colleagues first applied an ssOligo donor to produce knockin mice carrying SNVs (Meyer et al. 2012). They generated a knockin mouse carrying several SNVs including a missense mutation in the *Rab38* gene by co-injecting ZFN mRNAs with a 144-mer ssOligo containing seven substitutions into one-cell mouse embryos. They obtained one partially targeted mutant from 60 newborns, an efficiency of 1.7 %. This work clearly reveals the enormous potential of ssOligos for the replacement of conventional gene-targeting vectors in in vivo genome editing, a technology that should greatly facilitate the rapid production of knockin mice.

The first knockin mouse by TALEN-mediated in vivo genome editing was also achieved by Kühn and colleagues in early 2013 (Wefers et al. 2013). They generated knockin mice carrying *Rab38* missense SNV G19V by TALENs using an ssOligo. They first prepared TALENs targeting the same region of the *Rab38* gene previously targeted by themselves using ZFNs; they found that the activity of the TALENs was approximately twice that of ZFNs. They then co-injected TALENs and an ssOligo into one-cell mouse embryos and obtained one founder mouse carrying a partially targeted G19V allele from 117 newborns (an efficiency of 0.9 %).

Because TALENs are easier to construct as compared with ZFNs, knockin mice can be rapidly generated using a combination of TALENs and ssOligos. However, the relatively low knockin efficiency of TALENs is a bottleneck that limits the broad application of the method. Recently, they reported improved knockin efficiencies of up to 8 %, using TALEN mRNAs transcribed from plasmids containing a poly A tail, which may increase mRNA stability and translation efficiency (Panda et al. 2013).

Jones and Meisler (2014) reported the production of knockin mice carrying a missense mutation in *Scn8a,* which encodes the neuronal sodium channel Nav1.6 (the mutation causes human epilepsy), by TALEN-mediated in vivo genome editing. They used a conventional targeting vector as a knockin donor and obtained correctly targeted newborns with 7 % efficiency. *Scn8a* is a member of a sodium channel family with nine paralogs sharing high DNA sequence identity. To investigate the specificity of their TALENs, the authors analyzed candidate regions of *Scn5a* and Scn4a, with 3 and 4 mismatches with *Scn8a* TALEN binding sequences, respectively. Although they found no off-target modifications in *Scn5a* or Scn4a in their 5 correctly targeted knockin mice, surprisingly, they found several off-target modifications in both genes in NHEJ-mediated knockout mice. This work suggests the power of TALEN-mediated in vivo genome editing and the considerable problem of off-target effects, a point to be discussed later.

Wiles and colleagues systematically investigated the optimal conditions for TALEN-mediated in vivo genome editing with ssOligos (Low et al. 2014). They chose as a model the mouse *Crb1rd8* gene mutation, which is present in many inbred laboratory strains derived from C57BL/6N and leads to retinal external limiting membrane fragmentation and outer retinal dysplasia. They extensively compared TALEN mRNA concentrations (10, 25, and 50 ng/μL), and ssODN length (200- or 52-mer), strand orientation (sense or antisense), and concentration (0.3, 1, 2, 6 ng/μL). As results, they showed that higher concentrations of TALEN mRNA and longer sense ssODNs improved knockin efficiency. Under the best conditions (50 ng/μL TALEN mRNA, 200-mer sense ssODN), the knockin efficiency was 27 % of newborns. This work contributes useful fundamental information to the research community and accelerates knockin mouse production by in vivo genome editing.

Although TALEN-mediated in vivo genome editing is improving rapidly, the efficiency of knockin is still low compared to that of knockout mouse production. A key step to improve knockin efficiency is TALEN activity. Recently, Yamamoto and colleagues developed a novel TALEN platform, named Platinum TALEN, which shows extremely high efficiency (Sakuma et al. 2013). We focused on *Slc1a3*, a glial glutamate transporter, in knockout mice has been reported as the first model for normal tension glaucoma (Watase et al. 1998; Harada et al. 1998, 2007; Bai et al. 2013a, b; Namekata et al. 2013). We recently discovered deleterious rare missense SNVs in *SLC1A3*, a human ortholog of *Slc1a3*, in patients with glaucoma (Yanagisawa and Aida et al. unpublished observation). To test whether the causality of these rare SNVs for glaucoma pathophysiology in mice, we generated knockin mice with these SNVs in the *Slc1a3* gene. We co-injected highly active Platinum TALENs targeting *Slc1a3* into one-cell mouse embryos with ssOligos carrying each SNV. We obtained several germline-competent knockin founders with a targeting efficiency of approximately 20 % and there were no off-target effects (Aida et al., unpublished observation). Because a single microinjection is sufficient to produce several knockin founders, Platinum TALEN technology provides a fast and efficient approach for producing genetic mouse models that reproduce the disease-associated SNVs of complex diseases. When the Platinum TALEN is used under the optimal conditions reported by Low et al. (2014), it is expected that knockin efficiency will be further improved.

Mice carrying a large functional gene cassette such as fluorescent reporter proteins, Cre recombinase and Lox-flanked sequences are the fundamental tools for basic and pathophysiological biology. In contrast to the situation with SNV, there are few reports of the production of knockin mouse carrying a large cassette by in vivo genome editing. Beyer and colleagues reported the first knockin mice carrying a cassette for conditional targeting consisting of EGFP, two pairs of Lox sequences known as FLEX, and genomic fragments including two exons of the *Satb1* gene (Sommer et al. 2014). They identified a precisely targeted knockin mouse from 11 newborns under optimal conditions and confirmed that the conditional targeting of the cassette derived from the knockin mouse worked as expected. This work showed that knockin mouse production by TALEN-mediated in vivo genome editing is not limited to the insertion of SNVs with ssOligos to generate conditionally targeted alleles.

Thus, knockin mice carrying not only an SNV but also a large functional gene cassette can now be created within a month using TALEN-mediated in vivo genome editing with high efficiency. In summary, almost everything achieved by conventional ES cell-based gene targeting can now be performed by TALEN-mediated in vivo genome editing technologies. Further, this revolutionary technology allows previously impossible achievements, such as ultra-rapid production, biallelic targeting in F_0 mice, and multiplexing, making TALEN-mediated in vivo genome editing the first choice for gene targeting (Fig. 11.1).

11.5 Off-Target Effects

Off-target effects, which involve nonspecific recognition and digestion in non-targeted regions by ZFNs, TALENs, and the CRISPR/Cas9 system, is the foremost problem of genome editing and has been extensively discussed in the field. Compared to ZFNs, TALENs produce only minimal off-target effects (in less than a tenth), even at highly similar nonspecific target sites with only two mismatches in the TALEN recognition sequence in human cells (Mussolino et al. 2011).

Researchers at Cellectis, a leading supplier of custom-made TALENs, comprehensively analyzed the specificity of TALENs (Juillerat et al. 2014). By computational off-target site analysis of more than 15,500 putative TALENs against the human genome, they found that nearly two-thirds of these TALENs had potential off-target sites with more than 4 mismatches that strongly reduced the binding capacity of TALENs. For the experimental confirmation of TALEN specificity, they chose six TALENs that had at least one potential off-target site containing one to four mismatches and investigated their DNA modification activities in human cell lines. They analyzed potential off-target sites by deep sequencing and found that off-target modifications were detected at only a few loci, with extremely low modification efficiency.

Liu and colleagues performed unbiased prediction of potential off-target sequences of two TALENs using in vitro selection and high-throughput sequencing (Guilinger et al. 2014). They found 76 potential off-target sites in the human genome, 16 of which, surprisingly, were modified by TALENs in human cells. However, when they used engineered TALENs by substitution of several amino acids, off-target effects were tenfold lower than those of standard TALENs in human cells.

In contrast to the in vitro situation using human cell lines, off-target effects appear to be much rarer in TALEN-mediated in vivo genome editing. I previously reviewed off-target effects in mutant mice generated by TALEN-mediated in vivo genome editing described in three reports, and found no off-target effects in 15 potential off-target sites containing only one mismatch for four TALEN pairs (Aida et al. 2014; Sung et al. 2013; Panda et al. 2013; Qiu et al. 2013). Lower off-target effects in mutant mice produced by in vivo genome editing compared to those in human cell lines have also been reported for CRISPR/Cas9 (Yang et al. 2013). One possible explanation is differences in properties of fertilized eggs and transformed human cell lines, which may consist of heterogeneous populations with different and increased mutagenic possibility. Another is the means of delivery: short-lived RNA for fertilized embryos, and stable plasmid DNA, which shows longer and higher expression, for human cell lines. Thus, the specificity of in vivo genome editing in fertilized mouse eggs is higher than that in human cell lines, and useful for mutant mouse production. However, further extensive characterization of off-target effects in TALEN-mediated in vivo genome editing should be performed.

11.6 Applications

The mouse is already in an impregnable position in the field of gene targeting in mammals, and thousands of various knockout and knockin mice have been produced. However, the ease and speed of TALEN-mediated in vivo genome editing further accelerates and expands the utility and application of genetically targeted mice. In the conventional ES cell-based gene targeting method, it takes at least a year to obtain a homozygous mutant at best (Aida et al. 2014). Also, it is common to spend a year or more obtaining germline-competent chimeric founders. However, TALEN-mediated in vivo genome editing is revolutionizing these complex and long processes and enables the ultra-rapid production of precisely targeted homozygous mice within a month (Fig. 11.1; Sung et al. 2013; Qiu et al. 2013; Wu et al. 2013; Liu et al. 2014; Li et al. 2014). Thus, now researchers can freely and rapidly generate gene-targeted mice as they handle cell lines.

One promising application of this groundbreaking technology is the functional characterization of millions of common and rare human SNVs discovered in GWAS, and exome- and whole genome sequencing in vivo by generating "humanized" knockin mice carrying equivalent variants. Because the vast majority of rare SNVs are missense, synonymous, intronic, or intergenic, instead of gene-disrupting nonsense, splice site, or frameshift variants, they are equivalent to "knockout" alleles (Veltman and Brunner 2012), and the production of knockin mouse models carrying such SNVs is a much better approach than the production of conventional knockout mouse models. Jones and Meisler (2014) described a good example of this application. By TALEN-mediated in vivo genome editing, they generated knockin mice carrying a de novo missense mutation N1768D in *Scn8a* discovered by exome sequencing of patients with epileptic encephalopathy or intellectual disability. Interestingly, the authors noted that their N1768D/+heterozygous mice exhibited the key clinical features of patients with epileptic encephalopathy: behavioral abnormalities, seizures, and sudden, unexpected death. Thus, the ultra-rapid production of precisely modified mouse models by TALEN-mediated in vivo genome editing drastically accelerates the functional investigation of rare SNVs in human genetic diseases.

TALEN-mediated genome editing technology also accelerates functional research on common SNVs in regulatory noncoding regions catalogued by the ENCODE (Encyclopedia of DNA Elements) project and GWAS studies (ENCODE Project Consortium et al. 2012; Maurano et al. 2012). Orkin and colleagues found common SNVs associated with fetal hemoglobin level identified by GWAS located in the noncoding erythroid enhancer region of *BCL11A* (Bauer et al. 2013). They showed that the SNV disrupts a motif sequence for the binding of a transcription factor, resulting in reduced *BCL11A* expression and elevated fetal hemoglobin level. Further, they showed by generating a transgenic mouse carrying the human enhancer that the noncoding sequence functions as a developmental stage–specific, lineage-restricted enhancer in vivo. To investigate directly whether the enhancer regulates *BCL11A* expression, they identified a 10-kb orthologous enhancer signature in the noncoding region of the mouse *Bcl11a* and deleted the region using two pairs of TALENs in a mouse erythroleukemia cell line. Surprisingly, biallelic excision of the region by TALENs profoundly reduced *Bcl11a* transcript and completely removed BLC11A protein. This excellent work suggests that TALEN-mediated genome editing is an invaluable tool for functional genomics and potentially for cell-specific gene therapy. Although they deleted enhancer regions from a mouse cell line in this study, it is very interesting to generate, by using TALEN-mediated in vivo genome editing, knockin mice carrying GWAS-associated human SNVs in the orthologous enhancer, and investigate those phenotypes in vivo (Sur et al. 2012; Kamberov et al. 2013).

Genome editing technology provides the potential for application of precise gene targeting not only in mouse but also in higher species such as in nonhuman primates, to model human development and disease and investigate higher brain function. Recently, two teams applied TALEN-mediated in vivo genome editing to rhesus and cynomolgus monkeys to target *MeCP2,* a causal gene for Rett syndrome (Liu et al. 2014a, c). Male monkeys carrying *MeCP2* mutations died at midgestation (Liu et al. 2014a) or soon after birth (Liu et al. 2014c), consistent with male embryonic lethality in humans carrying *MeCP2* mutations. In contrast, female newborn monkeys carrying *MeCP2* mutations were obtained, consistent with the observation that patients with Rett syndrome are female, with no detectable off-target effects. These works provide proof of concept for gene targeting in non-human primates and opportunities to investigate higher brain function and develop novel therapeutics for human diseases that are difficult to study in mice.

TALEN consists of a TALE domain for specific DNA binding and an effector domain, which is a FokI nuclease domain for introducing a double-strand break into target DNA. Thus, in addition to targeted genome editing, TALEs can be applied for targeted genome regulation, and for visualization by exchanging the effector domain with other functional domains such as transcriptional and epigenetic effector domains or fluorescent proteins (Hsu and Zhang 2012). When a TALE is engineered to fuse with a transcriptional activator domain, the specific target gene can be specifically activated. Zhang and colleagues showed TALE-mediated specific transcriptional and epigenetic control in the adult brain of freely moving mice in vivo by a combination of genome editing with optogenetics, a high-precision spatiotemporal control of many cellular functions using light-responsive proteins (Konermann et al. 2013). They engineered a TALE to fuse with CIB1 protein, which physically binds, under blue light, with its interacting partner light-sensitive CRY2 protein fused with a VP64 transcriptional activator domain, leading to transcription of a target gene. They engineered TALEs to target the *Grm2* gene, packaged these components into viral vectors (adeno-associated viruses, AAVs), and delivered AAVs into adult brains. By blue light stimulation through a fiber optic cannula system, the expression level of endogenous *Grm2* was increased by more than twice. Further, they engineered the CRY2 to fuse with histone deacetylases (HDACs), histone methyltransferases (HMTs), histone deacetylases, histone acetyltransferase inhibitors, HDACs, and HMT recruiting proteins. When they tested these epigenetic TALEs with *Grm2* in mouse neural cells, the expression levels of *Grm2* were suppressed via histone modifications.

TALE can also be engineered with fluorescent proteins and be used as fluorescent probes to visualize specific chromosomal configurations in mouse in vivo. Torres-Padilla and colleagues reported TALE-mediated fluorescent genomic visualization of endogenous repetitive mouse genome sequences and showed nuclear positioning and chromatin dynamics in living mouse embryos (Miyanari et al. 2013). They engineered TALE to target mouse major satellite repeats, highly abundant tandem repeats in pericentromeric regions, and fused it with monomeric GFP mClover. Using this fluorescent TALE probe, they successfully visualized the replication and proliferation of major satellite repeats in living mouse embryos throughout development.

These advanced approaches suggest that TALE is a highly generalizable, robust and versatile platform for the manipulation of endogenous mammalian genomes and has unlimited potential for broad application. Overall, TALE and TALEN-mediated in vivo genome editing are revolutionary technologies and dramatically facilitate our understanding of in vivo gene functions in health and disease.

Acknowledgements This work was supported in part by Strategic Research Program for Brain Sciences and Grant-in-Aid for Science Research from Ministry of Education, Culture, Sports, Science and Technology of Japan, and a grant from Medical Research Institute, Tokyo Medical and Dental University. I thank K Tanaka, R Imahashi, H Ishikubo, and T Usami (Tokyo Medical and Dental University), and T Yamamoto, and T Sakuma (Hiroshima University) for technical support and useful discussions.

References

Aida T, Ito Y, Takahashi YK, Tanaka K (2012) Overstimulation of NMDA receptors impairs early brain development in vivo. PLoS One 7:e36853

Aida T, Imahashi R, Tanaka K (2014) Translating human genetics into mouse: the impact of ultra-rapid in vivo genome editing. Dev Growth Differ 56:34–45

Anonymous (2012) Method of the Year 2011. Nat Methods 9:1

1000 Genomes Project Consortium, Abecasis GR, Auton A, Brooks LD, DePristo MA, Durbin RM, Handsaker RE, Kang HM, Marth GT, McVean GA (2012) An integrated map of genetic variation from 1,092 human genomes. Nature 491:56–65

Bai N, Aida T, Yanagisawa M, Katou S, Sakimura K, Mishina M, Tanaka K (2013a) NMDA receptor subunits have different roles in NMDA-induced neurotoxicity in the retina. Mol Brain 6:34

Bai N, Hayashi H, Aida T, Namekata K, Harada T, Mishina M, Tanaka K (2013b) Dock3 interaction with a glutamate-receptor NR2D subunit protects neurons from excitotoxicity. Mol Brain 6:22

Bauer DE, Kamran SC, Lessard S, Xu J, Fujiwara Y, Lin C, Shao Z, Canver MC, Smith EC, Pinello L, Sabo PJ, Vierstra J, Voit RA, Yuan GC, Porteus MH, Stamatoyannopoulos JA, Lettre G, Orkin SH (2013) An erythroid enhancer of BCL11A subject to genetic variation determines fetal hemoglobin level. Science 342:253–257

Brinster RL, Braun RE, Lo D, Avarbock MR, Oram F, Palmiter RD (1989) Targeted correction of a major histocompatibility class II E alpha gene by DNA microinjected into mouse eggs. Proc Natl Acad Sci U S A 86:7087–7091

Capecchi MR (2005) Gene targeting in mice: functional analysis of the mammalian genome for the twenty-first century. Nat Rev Genet 6:507–512

Carbery ID, Ji D, Harrington A, Brown V, Weinstein EJ, Liaw L, Cui X (2010) Targeted genome modification in mice using zinc-finger nucleases. Genetics 186:451–459

Carroll D (2014) Genome engineering with targetable nucleases. Annu Rev Biochem 83:409–439

Chen F, Pruett-Miller SM, Huang Y, Gjoka M, Duda K, Taunton J, Collingwood TN, Frodin M, Davis GD (2011) High-frequency genome editing using ssDNA oligonucleotides with zinc-finger nucleases. Nat Methods 8:753–755

Cui X, Ji D, Fisher DA, Wu Y, Briner DM, Weinstein EJ (2011) Targeted integration in rat and mouse embryos with zinc-finger nucleases. Nat Biotechnol 29:64–67

Davies B, Davies G, Preece C, Puliyadi R, Szumska D, Bhattacharya S (2013) Site specific mutation of the Zic2 locus by microinjection of TALEN mRNA in mouse CD1, C3H and C57BL/6 J oocytes. PLoS One 8:e60216

Echeverri CJ, Beachy PA, Baum B, Boutros M, Buchholz F, Chanda SK, Downward J, Ellenberg J, Fraser AG, Hacohen N, Hahn WC, Jackson AL, Kiger A, Linsley PS, Lum L, Ma Y, Mathey-Prévôt B, Root DE, Sabatini DM, Taipale J, Perrimon N, Bernards R (2006) Minimizing the risk of reporting false positives in large-scale RNAi screens. Nat Methods 3:777–779

ENCODE Project Consortium, Bernstein BE, Birney E, Dunham I, Green ED, Gunter C, Snyder M (2012) An integrated encyclopedia of DNA elements in the human genome. Nature 489:57–74

Geurts AM, Cost GJ, Freyvert Y, Zeitler B, Miller JC, Choi VM, Jenkins SS, Wood A, Cui X, Meng X, Vincent A, Lam S, Michalkiewicz M, Schilling R, Foeckler J, Kalloway S, Weiler H, Ménoret S,

Anegon I, Davis GD, Zhang L, Rebar EJ, Gregory PD, Urnov FD, Jacob HJ, Buelow R (2009) Knockout rats via embryo microinjection of zinc-finger nucleases. Science 325:433
Guilinger JP, Pattanayak V, Reyon D, Tsai SQ, Sander JD, Joung JK, Liu DR (2014) Broad specificity profiling of TALENs results in engineered nucleases with improved DNA-cleavage specificity. Nat Methods 11:429–435
Harada T, Harada C, Watanabe M, Inoue Y, Sakagawa T, Nakayama N, Sasaki S, Okuyama S, Watase K, Wada K, Tanaka K (1998) Functions of the two glutamate transporters GLAST and GLT-1 in the retina. Proc Natl Acad Sci U S A 95:4663–4666
Harada T, Harada C, Nakamura K, Quah HM, Okumura A, Namekata K, Saeki T, Aihara M, Yoshida H, Mitani A, Tanaka K (2007) The potential role of glutamate transporters in the pathogenesis of normal tension glaucoma. J Clin Invest 117:1763–1770
Hauschild-Quintern J, Petersen B, Cost GJ, Niemann H (2013) Gene knockout and knockin by zinc-finger nucleases: current status and perspectives. Cell Mol Life Sci 70:2969–2983
Hsu PD, Zhang F (2012) Dissecting neural function using targeted genome engineering technologies. ACS Chem Neurosci 3:603–610
Hsu PD, Lander ES, Zhang F (2014) Development and applications of CRISPR-Cas9 for genome engineering. Cell 157:1262–1278
Jones JM, Meisler MH (2014) Modeling human epilepsy by TALEN targeting of mouse sodium channel Scn8a. Genesis 52:141–148
Joung JK, Sander JD (2013) TALENs: a widely applicable technology for targeted genome editing. Nat Rev Mol Cell Biol 14:49–55
Juillerat A, Dubois G, Valton J, Thomas S, Stella S, Maréchal A, Langevin S, Benomari N, Bertonati C, Silva GH, Daboussi F, Epinat JC, Montoya G, Duclert A, Duchateau P (2014) Comprehensive analysis of the specificity of transcription activator-like effector nucleases. Nucleic Acids Res 42:5390–5402
Kamberov YG, Wang S, Tan J, Gerbault P, Wark A, Tan L, Yang Y, Li S, Tang K, Chen H, Powell A, Itan Y, Fuller D, Lohmueller J, Mao J, Schachar A, Paymer M, Hostetter E, Byrne E, Burnett M, McMahon AP, Thomas MG, Lieberman DE, Jin L, Tabin CJ, Morgan BA, Sabeti PC (2013) Modeling recent human evolution in mice by expression of a selected EDAR variant. Cell 152:691–702
Kato T, Miyata K, Sonobe M, Yamashita S, Tamano M, Miura K, Kanai Y, Miyamoto S, Sakuma T, Yamamoto T, Inui M, Kikusui T, Asahara H, Takada S (2013) Production of Sry knockout mouse using TALEN via oocyte injection. Sci Rep 3:3136
Kitajima S, Takagi A, Inoue T, Saga Y (2000) MesP1 and MesP2 are essential for the development of cardiac mesoderm. Development 127:3215–3226
Konermann S, Brigham MD, Trevino AE, Hsu PD, Heidenreich M, Cong L, Platt RJ, Scott DA, Church GM, Zhang F (2013) Optical control of mammalian endogenous transcription and epigenetic states. Nature 500:472–476
Li C, Qi R, Singleterry R, Hyle J, Balch A, Li X, Sublett J, Berns H, Valentine M, Valentine V, Sherr CJ (2014) Simultaneous gene editing by injection of mRNAs encoding transcription activator-like effector nucleases into mouse zygotes. Mol Cell Biol 34:1649–1658
Liu H, Chen Y, Niu Y, Zhang K, Kang Y, Ge W, Liu X, Zhao E, Wang C, Lin S, Jing B, Si C, Lin Q, Chen X, Lin H, Pu X, Wang Y, Qin B, Wang F, Wang H, Si W, Zhou J, Tan T, Li T, Ji S, Xue Z, Luo Y, Cheng L, Zhou Q, Li S, Sun YE, Ji W (2014a) TALEN-mediated gene mutagenesis in rhesus and cynomolgus monkeys. Cell Stem Cell 14:323–328
Liu Y, Lv X, Tan R, Liu T, Chen T, Li M, Liu Y, Nie F, Wang X, Zhou P, Chen M, Zhou Q (2014b) A modified TALEN-based strategy for rapidly and efficiently generating knockout mice for kidney development studies. PLoS One 9:e84893
Liu Z, Zhou X, Zhu Y, Chen ZF, Yu B, Wang Y, Zhang CC, Nie YH, Sang X, Cai YJ, Zhang YF, Zhang C, Zhou WH, Sun Q, Qiu Z (2014c) Generation of a monkey with MECP2 mutations by TALEN-based gene targeting. Neurosci Bull 30:381–386
Low BE, Krebs MP, Joung JK, Tsai SQ, Nishina PM, Wiles MV (2014) Correction of the Crb1rd8 allele and retinal phenotype in C57BL/6N mice via TALEN-mediated homology-directed repair. Invest Ophthalmol Vis Sci 55:387–395

Matsugami TR, Tanemura K, Mieda M, Nakatomi R, Yamada K, Kondo T, Ogawa M, Obata K, Watanabe M, Hashikawa T, Tanaka K (2006) Indispensability of the glutamate transporters GLAST and GLT1 to brain development. Proc Natl Acad Sci U S A 103:12161–12166

Maurano MT, Humbert R, Rynes E, Thurman RE, Haugen E, Wang H, Reynolds AP, Sandstrom R, Qu H, Brody J, Shafer A, Neri F, Lee K, Kutyavin T, Stehling-Sun S, Johnson AK, Canfield TK, Giste E, Diegel M, Bates D, Hansen RS, Neph S, Sabo PJ, Heimfeld S, Raubitschek A, Ziegler S, Cotsapas C, Sotoodehnia N, Glass I, Sunyaev SR, Kaul R, Stamatoyannopoulos JA (2012) Systematic localization of common disease-associated variation in regulatory DNA. Science 337:1190–1195

Menke DB (2013) Engineering subtle targeted mutations into the mouse genome. Genesis 51:605–618

Meyer M, de Angelis MH, Wurst W, Kühn R (2010) Gene targeting by homologous recombination in mouse zygotes mediated by zinc-finger nucleases. Proc Natl Acad Sci U S A 107:15022–15026

Meyer M, Ortiz O, Hrabé de Angelis M, Wurst W, Kühn R (2012) Modeling disease mutations by gene targeting in one-cell mouse embryos. Proc Natl Acad Sci U S A 109:9354–9359

Miyanari Y, Ziegler-Birling C, Torres-Padilla ME (2013) Live visualization of chromatin dynamics with fluorescent TALEs. Nat Struct Mol Biol 20:1321–1324

Mussolino C, Morbitzer R, Lütge F, Dannemann N, Lahaye T, Cathomen T (2011) A novel TALE nuclease scaffold enables high genome editing activity in combination with low toxicity. Nucleic Acids Res 39:9283–9293

Namekata K, Kimura A, Kawamura K, Guo X, Harada C, Tanaka K, Harada T (2013) Dock3 attenuates neural cell death due to NMDA neurotoxicity and oxidative stress in a mouse model of normal tension glaucoma. Cell Death Differ 20:1250–1256

Panda SK, Wefers B, Ortiz O, Floss T, Schmid B, Haass C, Wurst W, Kühn R (2013) Highly efficient targeted mutagenesis in mice using TALENs. Genetics 195:703–713

Qiu Z, Liu M, Chen Z, Shao Y, Pan H, Wei G, Yu C, Zhang L, Li X, Wang P, Fan HY, Du B, Liu B, Liu M, Li D (2013) High-efficiency and heritable gene targeting in mouse by transcription activator-like effector nucleases. Nucleic Acids Res 41:e120

Raychaudhuri S (2011) Mapping rare and common causal alleles for complex human diseases. Cell 147:57–69

Saito T, Matsuba Y, Mihira N, Takano J, Nilsson P, Itohara S, Iwata N, Saido TC (2014) Single App knock-in mouse models of Alzheimer's disease. Nat Neurosci 17:661–663

Sakuma T, Ochiai H, Kaneko T, Mashimo T, Tokumasu D, Sakane Y, Suzuki K, Miyamoto T, Sakamoto N, Matsuura S, Yamamoto T (2013) Repeating pattern of non-RVD variations in DNA-binding modules enhances TALEN activity. Sci Rep 3:3379

Skarnes WC, Rosen B, West AP, Koutsourakis M, Bushell W, Iyer V, Mujica AO, Thomas M, Harrow J, Cox T, Jackson D, Severin J, Biggs P, Fu J, Nefedov M, de Jong PJ, Stewart AF, Bradley A (2011) A conditional knockout resource for the genome-wide study of mouse gene function. Nature 474:337–342

Sommer D, Peters A, Wirtz T, Mai M, Ackermann J, Thabet Y, Schmidt J, Weighardt H, Wunderlich FT, Degen J, Schultze JL, Beyer M (2014) Efficient genome engineering by targeted homologous recombination in mouse embryos using transcription activator-like effector nucleases. Nat Commun 5:3045

Sung YH, Baek IJ, Seong JK, Kim JS, Lee HW (2012) Mouse genetics: catalogue and scissors. BMB Rep 45:686–692

Sung YH, Baek IJ, Kim DH, Jeon J, Lee J, Lee K, Jeong D, Kim JS, Lee HW (2013) Knockout mice created by TALEN-mediated gene targeting. Nat Biotechnol 31:23–24

Sur IK, Hallikas O, Vähärautio A, Yan J, Turunen M, Enge M, Taipale M, Karhu A, Aaltonen LA, Taipale J (2012) Mice lacking a Myc enhancer that includes human SNP rs6983267 are resistant to intestinal tumors. Science 338:1360–1363

Tabuchi K, Blundell J, Etherton MR, Hammer RE, Liu X, Powell CM, Südhof TC (2007) A neuroligin-3 mutation implicated in autism increases inhibitory synaptic transmission in mice. Science 318:71–76

Takada S, Sato T, Ito Y, Yamashita S, Kato T, Kawasumi M, Kanai-Azuma M, Igarashi A, Kato T, Tamano M, Asahara H (2013) Targeted gene deletion of miRNAs in mice by TALEN system. PLoS One 8:e76004

Urnov FD, Rebar EJ, Holmes MC, Zhang HS, Gregory PD (2010) Genome editing with engineered zinc finger nucleases. Nat Rev Genet 11:636–646

Veltman JA, Brunner HG (2012) De novo mutations in human genetic disease. Nat Rev Genet 13:565–575

Wang H, Hu YC, Markoulaki S, Welstead GG, Cheng AW, Shivalila CS, Pyntikova T, Dadon DB, Voytas DF, Bogdanove AJ, Page DC, Jaenisch R (2013) TALEN-mediated editing of the mouse Y chromosome. Nat Biotechnol 31:530–532

Watase K, Hashimoto K, Kano M, Yamada K, Watanabe M, Inoue Y, Okuyama S, Sakagawa T, Ogawa S, Kawashima N, Hori S, Takimoto M, Wada K, Tanaka K (1998) Motor discoordination and increased susceptibility to cerebellar injury in GLAST mutant mice. Eur J Neurosci 10:976–988

Wefers B, Panda SK, Ortiz O, Brandl C, Hensler S, Hansen J, Wurst W, Kühn R (2013) Generation of targeted mouse mutants by embryo microinjection of TALEN mRNA. Nat Protoc 8:2355–2379

Wu J, Huang Z, Ren J, Zhang Z, He P, Li Y, Ma J, Chen W, Zhang Y, Zhou X, Yang Z, Wu SQ, Chen L, Han J (2013) Mlkl knockout mice demonstrate the indispensable role of Mlkl in necroptosis. Cell Res 23:994–1006

Yang H, Wang H, Shivalila CS, Cheng AW, Shi L, Jaenisch R (2013) One-step generation of mice carrying reporter and conditional alleles by CRISPR/Cas-mediated genome engineering. Cell 154:1370–1379

Chapter 12
Engineered Nucleases Lead to Genome Editing Revolution in Rats

Kazuto Yoshimi, Takehito Kaneko, Birger Voigt, and Tomoji Mashimo

Abstract Sequence-specific endonucleases, such as zinc finger nucleases (ZFNs), transcription activator-like effector nucleases (TALENs), and clustered regularly interspaced short palindromic repeats (CRISPR)/CRISPR-associated (Cas), allow simple generation of genetically modified animals. We utilized ZFNs and TALENs to generate several knockout rat models of human diseases. Furthermore, CRISPR/Cas9 used in conjunction with single-stranded oligodeoxyribonucleotides enabled us to generate several types of targeted knockin animal, such as single nucleotide polymorphism substitution, short DNA fragment integration, and large DNA fragment elimination. These powerful technologies have opened new doors for genome manipulation that enable the engineering of animal models of human disease and potential therapeutic applications.

Keywords Allele-specific gene editing • Genetically modified animals • Knockin • Rats • Single-stranded oligodeoxyribonucleotides

12.1 Genome Engineering Technologies in Rats

The laboratory rat (*Rattus norvegicus*) has been widely used to generate in vivo models for studying human diseases such as hypertension (Yamori 1991), diabetes (Chappel and Chappel 1983; Kawano et al. 1994), epilepsy (Serikawa and Yamada 1986; Mashimo et al. 2010a), inflammation (Kleinau et al. 1994) and cancer (Mori et al. 1994; Yoshimi et al. 2009). It is also used as a model for assessing the beneficial and toxicological effects of compounds and drugs. Because of its larger body size and ease of manipulation compared with the laboratory mouse (*Mus musculus*), it is very useful for longitudinal and challenging experiments that require surgical manipulation, such as transplantation or neurological research. Furthermore its physiological characteristics are similar to those of humans. These advantages have greatly contributed to the understanding of fundamental physiological mechanisms

K. Yoshimi • T. Kaneko • B. Voigt • T. Mashimo (✉)
Institute of Laboratory Animals, Graduate School of Medicine,
Kyoto University, Kyoto, Japan
e-mail: tmashimo@anim.med.kyoto-u.ac.jp

T. Yamamoto (ed.), *Targeted Genome Editing Using Site-Specific Nucleases*,
DOI 10.1007/978-4-431-55227-7_12

in humans as well as diseases mechanisms and the development of therapeutic agents and regimens.

The development of gene targeting in mice through targeted homologous recombination in embryonic stem (ES) cells began in the end of 1980s (Capecchi 1989). Knockout (KO) mice, in which a null allele of a gene is engineered, can directly contribute to reveal the in vivo function of specific genes. Furthermore, knockin (KI) mice, in which genes are added or modified, and conditional knockout/in mice in which gene activity can be spatially and/or temporally controlled, are also widely used to avoid early lethal phenotypes and to address biological questions with great accuracy (Lewandoski 2001; Stanford et al. 2001). These ES cell-mediated gene targeting technologies have become critical tools for understanding gene functions, including the genetic basis of human diseases.

Until recently it was difficult to produce mammalian KO animals using gene targeting technologies in any species other than mouse because only mouse germline-competent ES cells were available. However, this situation changed with the availability of newly developed gene targeting technologies that use engineered nucleases or "gene scissors" (Table 12.1). These engineered nucleases, include zinc finger nucleases (ZFNs), transcription activator-like effector nucleases (TALENs), or clustered regularly interspaced short palindromic repeats (CRISPR)/CRISPR-associated (Cas). They are highly effective when microinjected into embryos for the generation of targeted KO mice and rats (Fig. 12.1) (Remy et al. 2010; Jacob et al.

Table 12.1 Characteristics of genome editing technologies in rats

Genome editing technologies	Characterizations	References
ES cells	12 to 18 months for generating KO rats	Tong et al. (2010); Yamamoto et al. (2012)
	Limited availability of strain specific ES cells	
	Sophisticated manipulation requirement for generating chimeric rats	
ZFN	4 to 6 months for generating KO rats	Geurts et al. (2009); Mashimo et al. (2010b); Cui et al. (2011)
	Numerous gene-modified rats have been published	
	Complicated ZFN design/construction on limited target sequences	
	Patent related issues	
TALEN	2 to 6 months for generating KO rats	Tesson et al. (2011); Mashimo et al. (2013)
	Flexibility to design/construct TALENs across the whole genome	
	Library for generating target-specific TALENs is required	
CRISPR/Cas9	2 to 6 months for generating KO rats	Li et al. (2013a, b); Bassett et al. (2013); Ma et al. (2014a); Yoshimi et al. (2014)
	Highly flexible to design/construct gRNAs for target sequences across the whole genome	
	Easy construction of target-specific gRNAs at low cost	
	Simultaneous modification of multiple genes (multiplex editing)	

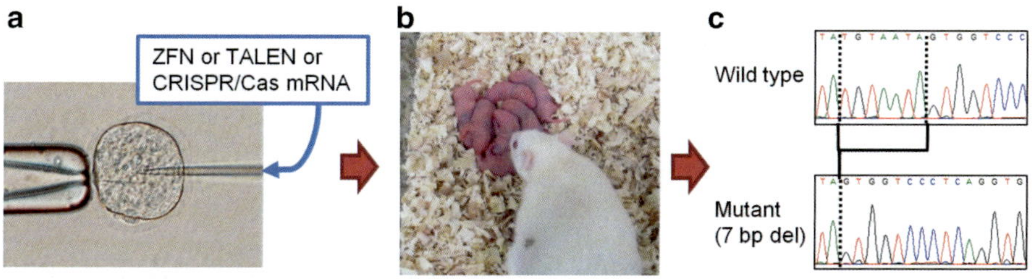

Fig. 12.1 Generation of genetically modified rats with genome editing technologies. (**a**) Microinjection of ZFN mRNA, TALEN mRNA or Cas9 mRNA and gRNA into pronuclear-stage rat embryos. (**b**) G0 pups derived from microinjected embryos three weeks after transfer to pseudopregnant females. (**c**) Detection of G0 founders carrying mutations at the target site by DNA sequencing analysis

2010; Mashimo 2014). In this review, we discuss the advantages and disadvantages of these site-specific nuclease technologies in relation to genetic analysis and gene manipulation in animals, especially rats.

12.2 Gene Editing with ZFNs in Rats

ZFNs are artificial fusion proteins that consist of an engineered DNA binding domain, made of tandem zinc-finger motifs with customized specificity, fused to a non-specific nuclease domain from the restriction endonuclease *FokI* (Bibikova et al. 2003; Porteus and Carroll 2005; Wu et al. 2007). *FokI* endonuclease can introduce a double-strand break (DSB) at the targeted locus following recognition of the specific DNA sequence by the zinc finger motifs. Although the DSB is usually repaired via non-homologous end joining (NHEJ), an arbitrary deletion or insertion of base pairs often occurs during the repair process. Consequently, repair by NHEJ is mutagenic and mostly results in a loss of function mutation (Bibikova et al. 2003; Porteus and Carroll 2005; Wu et al. 2007).

Numerous genetically-modified rats, generated using ZFN technology, have been reported since the first KO rat was reported in 2009 (Geurts et al. 2009; Mashimo et al. 2010b; Cui et al. 2011). We developed an interleukin-2 receptor gamma chain (*Il2rg*) KO rat (X-SCID) to investigate human X-linked severe combined immunodeficiency (X-SCID) (Mashimo et al. 2010b). We have also generated SCID rats that are deficient for the *Prkdc* gene, and F344-*scid Il2rg* (FSG) rats that are simultaneously deficient for both the *Prkdc* and *Il2rg* genes (Fig. 12.2) (Mashimo et al. 2012). In contrast to the "leaky" phenotype of the SCID mouse, where immunoglobulins such as IgG, are detected in the blood, SCID rats did not show such a leaky phenotype (Bosma et al. 1988). These SCID rats are available from the National Bio Resource Project for the Rat in Japan (NBRP-Rat: www.anim.med.kyoto-u.ac.jp/nbr). They demonstrate the applicability of the ZFN system for generating disease-related models and genetically modified hosts for xenotransplantation of human stem cells and tissues (Shultz et al. 2012).

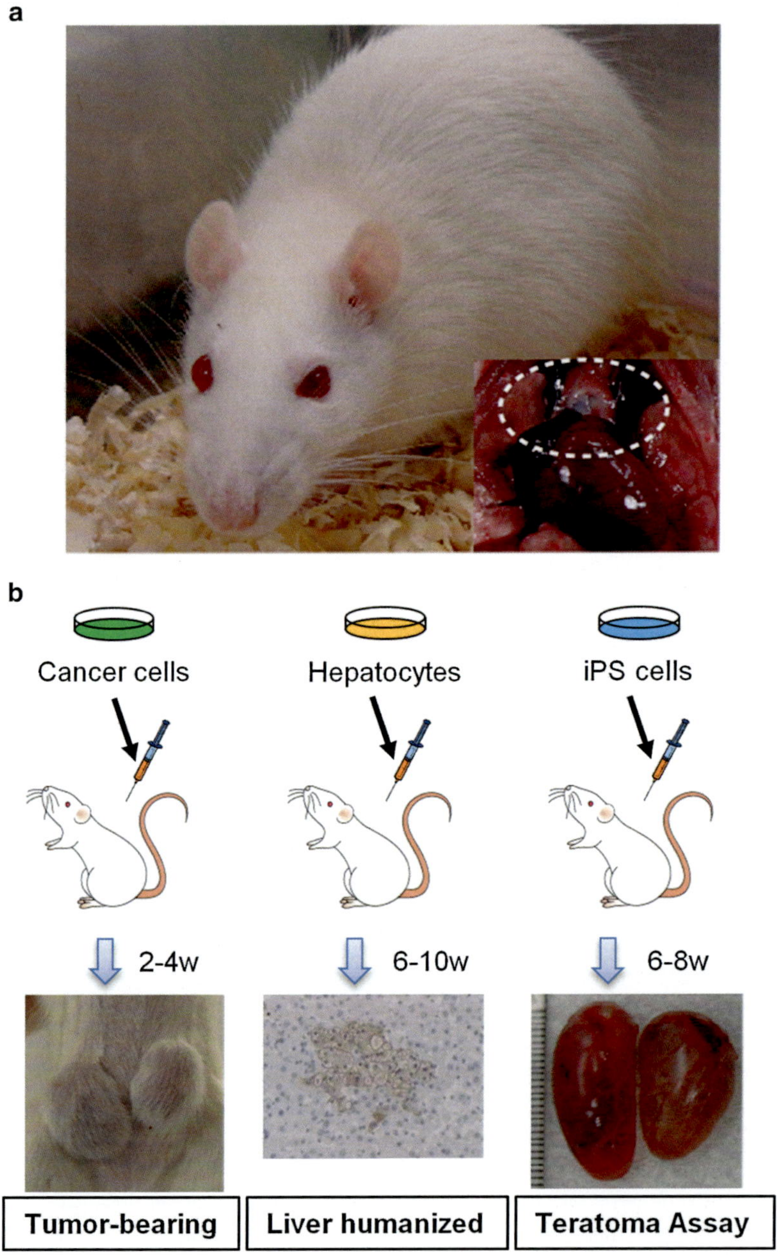

Fig. 12.2 Severe combined immunodeficiency (SCID) rats generated by ZFNs. (**a**) Immunodeficient rats show an atrophied thymus and immune cell deficiencies. (**b**) SCID rats can tolerate xenograft-experiments with human ovarian cancer cells, human hepatocytes and human induced pluripotent stem cells

12.3 Gene Editing with TALENs in Rats

Natural TAL effectors are potent virulence proteins from the pathogenic plant bacteria *Xanthomonas*. They can be injected into eukaryotic host cells where they function as transcription factors (Mussolino and Cathomen 2012; Joung and Sander 2013). TALENs are fusions of TAL effectors and the *FokI* nuclease and pairs of TALENs can bind and cleave DNA. While the targeted sequences recognized by ZF domains are limited for ZFNs, TAL effectors can recognize almost any sequence, except T at position 0. Simple and straightforward design and assembly strategies have been developed for rapid construction of TALENs, providing a cost-effective targeted nuclease platform (Christian et al. 2010; Hockemeyer et al. 2011; Sung et al. 2013; Wood et al. 2011).

Although TALEN technologies seem to have advantages over ZFNs, there are also some drawbacks that need to be addressed. For unknown reasons, the system appears to be less effective in rodent embryos. However, we recently showed that combined expression of exonuclease 1 (*Exo1*) with engineered site-specific TALENs provided highly efficient for the disruption of the albino (*Tyr*) gene in rat zygotes and in the production of *Tyr* knockout rats (Fig. 12.3) (Mashimo et al. 2013).

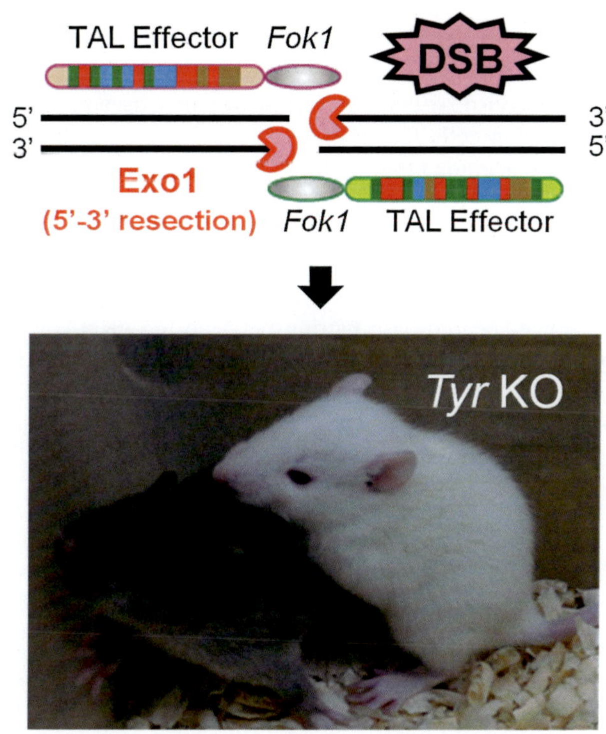

Fig. 12.3 Targeted KO rats generated by TALENs. Co-expression of exonuclease 1 (Exo1) can improve the efficiency of targeted KO mutagenesis induced by site-specific TALENs

The microinjection of TALENs with Exo1 is an easy and efficient method of generating gene knockouts using zygotes, which increases the range of species for which gene targeting technologies are available.

Several methods of constructing TALENs harboring different TALE scaffolds and repeat variants have been reported (Cermak et al. 2011; Hockemeyer et al. 2011; Huang et al. 2011; Joung and Sander 2013; Li et al. 2012). However, the structural differences among them remain unclear. Recently, Sakuma et al. showed that TALENs with periodically-patterned repeat variants harboring non-repeat-variable di-residue (non-RVD) variations (named Platinum TALENs) provide higher activities compared with TALENs without non-RVD variations (Sakuma et al. 2013). This Platinum Gate TALEN construction system has been deposited with Addgene as the 'TALEN Construction and Evaluation Accessory Pack' (www.addgene.org/TALEN/PlatinumGate). We applied Platinum TALENs to target the *Il2rg* gene in rat zygotes, and the mutant-generating efficiency was much higher than that of other TALENs (Tesson et al. 2011; Mashimo et al. 2013) and ZFNs (Mashimo et al. 2010b, 2012). Platinum TALENs have already been used in other species (Sakuma et al. 2013; Treen et al. 2014; Nakagawa et al. 2014) and have made a valuable contribution to genome editing research.

12.4 Genome Editing with CRISPR/Cas9 in Rats

The CRISPR/Cas9 system was first demonstrated as a gene-targeting technology in mammalian cells (Jinek et al. 2012; Cong et al. 2013; Mali et al. 2013b). Cas9 nuclease is guided to introduce DSBs at the targeted locus by an engineered single-guide RNA (gRNA) that recognizes the DNA sequence of the target. Therefore it is possible to direct Cas9 to any genomic locus simply by designing the appropriate synthetic gRNA. In contrast to ZFN and TALEN methods, which recognize targeted sequences by protein-DNA interaction, CRISPR/Cas9 can recognize targets by sequence-specific base pairing between the gRNA and the targeted site. Specific synthetic gRNA-expression plasmids can be easily and rapidly generated. This simple design and its targeting precision have made the CRISPR/Cas9 system extremely popular as a gene modification tool in a variety of cells and organisms (Cho et al. 2013; Fu et al. 2013; Hwang et al. 2013; Wang et al. 2013; Yang et al. 2013).

12.4.1 Knockout Rats with CRISPR/Cas9

Several studies have shown that the CRISPR/Cas9 system is an efficient tool for generating targeted KO rats (Li et al. 2013a, b; Ma et al. 2014a, b). We have also generated a targeted KO rat for the coat color gene, tyrosinase (*Tyr*), by microinjecting target-specific gRNA and Cas9 mRNA into fertilized eggs (Yoshimi et al. 2014). Crossing founders demonstrated that the CRISPR/Cas9-mediated mutations were faithfully transmitted to the next generation (Li et al. 2013a, b; Ma et al. 2014b). This system can also introduce multiple gene mutations in one step by microinjecting Cas9 mRNA and multiple gRNAs into rat zygotes. This is one of the significant advantages of this system (Cong et al. 2013; Wang et al. 2013; Ma et al. 2014a). Although it is possible to generate rats with multiple mutations by breeding together rats with single mutations, the process is costly and time-consuming. In addition it becomes harder to generate rats with multiple mutations in more closely linked genes. Multiplex gene modification by the CRISPR/Cas9 system enables the investigation of in vivo functions of tightly linked genes and gene clusters in a more rapid and cost effective way compared with the traditional ES-cell based approaches.

12.4.2 Specificity of CRISPR to Targeted Sites

gRNAs are designed to direct the Cas9 protein to a target site by recognizing 20 nucleotides near a 5′-NGG protospacer adjacent motif (PAM) sequence. Thus, altering the first 20 bases of a gRNA designed to correspond with a target site can introduce indel mutations in any DNA sequence of the form N20-NGG. Although the CRISPR/Cas9 system can generally introduce mutations at target sites reliably, several groups reported that the Cas9 nuclease can induce off-target mutagenesis in cell lines (Fu et al. 2013; Mali et al. 2013a; Hsu et al. 2013). These studies suggested that one to six mismatches are tolerated depending on the position and that bases 8–14 at the 3′ end of a target sequence is the critical region for recognition by gRNA. However, in contrast to cell-based experiments, most reports show that genetically-modified animals produced by RNA injection have low off-target frequency (Wang et al. 2013; Yang et al. 2013; Hu et al. 2013; Li et al. 2013a, b; Bassett et al. 2013; Friedland et al. 2013; Hwang et al. 2013). Furthermore, we have not detected any off-target site mutations in over thirty KO founder rats (Yoshimi et al. 2014). Further to these findings, off-target effects are likely to only play a minor role in an animal's phenotype because they will be lost in subsequent breeding.

In fact, an important characteristic of the CRISPR/Cas9 system is the high specificity of the bases that are 8–14 nucleotides from the PAM sequence that recognize target sites in the genome. We focused on a single nucleotide polymorphism (SNP) mutation to check whether the system can discriminate a single base within the region 8–14 bases from the PAM sequence. Albino rats such as Wistar and F344 strains carry a single SNP mutation 896G > A in exon 2 of the *Tyr* gene which causes to lack of pigmentation (Blaszczyk et al. 2005). We designed two gRNAs at the SNP site to specifically recognize both target sequences. We demonstrated that the CRISPR/Cas9 system can recognize single base pair differences and induce mutations at allele-specific target DNAs in embryos, suggesting that the system can facilitate allele-specific genome editing for heterogeneous strains or human/primate cell lines or stem cells that are normally genetically very heterogeneous (Yoshimi et al. 2014).

Even though the Cas9-induced off-target effects are a potential limitation for using the CRISPR/Cas9 system, search tools for target sequences with sophisticated algorithms that minimize off-target mutations are already available for many organisms (Hsu et al. 2013).

12.4.3 Knockin Rats with ssODNs

When Cas9 induces DSBs at targeted sites, homology directed repair (HDR) is stimulated by donor DNA templates such as double-stranded plasmids and single-stranded oligodeoxyribonucleotides (ssODNs). Targeted KI mice and rats have been generated via HDR-mediated genome editing by simply applying ZFN, TALEN or CRISPR/Cas9 architectures together with donor DNA templates (Cui et al. 2011; Brown et al. 2013; Ponce de Leon et al. 2014; Yang et al. 2013; Ma et al. 2014b). Although co-injecting DNA plasmids can introduce targeted insertions of several cassettes and reporter genes into the targeted sites, such events are characterized by a low efficiency of generating viable animals. Co-injection of ssODNs with Cas9 and gRNA can also introduce designed small mutations at target sites in mice with more than ten percent efficiency (Wang et al. 2013). We efficiently generated targeted KI mutations in rats with ssODNs and successfully recovered three distinct coat-color phenotypes: albino (*c*) a missense mutation in the *Tyr* gene, non-agouti (*a*) a 19 bp deletion in the *Asip* gene, and hooded (h) an integration of a 7,098 bp endogenous retroviral element in the *Kit* gene using the CRISPR/Cas9 platform (Fig. 12.4) (Yoshimi et al. 2014). The high efficiency and specificity of CRISPR/Cas9-mediated genome editing in rats prompted us to modify observable phenotypic traits and to create disease-causing mutations as therapeutic models of human diseases.

In the last decade, genome wide association studies have successfully linked numerous indel mutations and SNPs with major human diseases. The in vivo function of these variants should be tested to understand the mechanisms of these diseases and to develop treatments and means of prevention. Precise and accurate genome editing technologies can provide the necessary tools for generating KI rats carrying mutations equivalent to human-specific variants. Furthermore, modification of disease-causative DNA sequences in animals can pave the road for the therapy of genetic diseases in humans. Delivery of the CRISPR/Cas9 components to adult livers of genetically liver-injured mice corrected the mutation in hepatocytes and rescued their disease phenotype (Yin et al. 2014). This report suggests that in vivo genome editing with the CRISPR/Cas9 system has the potential to correct human genes to treat genetic diseases.

12.5 Advantages of Site-Specific Nuclease Technologies

For targeted gene manipulations in animal models, all artificial nuclease technologies (ZFN/TALEN/CRISPR) share the following advantages compared with conventional ES cell-based approaches (Table 12.1). (1) KO rats can be generated in a

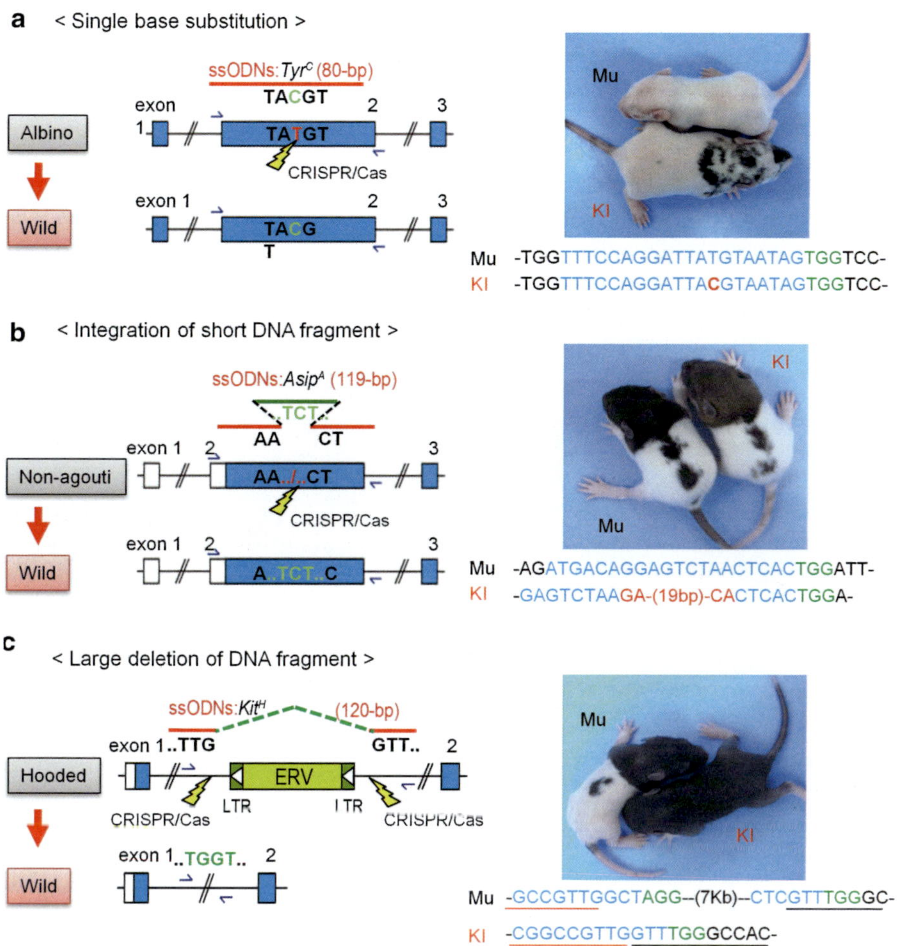

Fig. 12.4 Targeted KI rats with CRISPR/Cas9. Several types of targeted KI rats using single-stranded oligodeoxyribonucleotides (ssODNs). (**a**) The recessive albino phenotype was recovered to a wild-type coat-color by a single base substitution in the *Tyr* gene. (**b**) The recessive non-agouti phenotype was recovered to the agouti coat-color by a nineteen base insertion into the *Asip* gene. (**c**) The recessive hooded phenotype was recovered to the non-hooded coat-color by a 7 Kb deletion of an endogenous retrovirus element in the *Kit* gene

4–6-month timeframe and with an efficiency of more than 20 %. This is more favorable than the ES cell-based method for mice, which usually takes 12–18 months. Given the high rate of germline transmission, preliminary phenotypic analysis can be performed on G1 animals after intercrossing the initial G0 founders, thereby saving time and effort. (2) Gene targeting with artificial nucleases is not strain dependent and, accordingly, can be performed with any inbred strain. This means that existing strains can be directly employing for targeted gene disruption, thereby bypassing tedious and time-consuming backcrossing steps that generally take 2–3 years to complete. (3) The artificial nuclease technologies can be used to induce a wide variety of allelic changes covering small or large deletions or insertions. It is also feasible to use targeted KI technologies that have thus far been inaccessible owing to the lack of rat ES cells. In particular, the use of ssODNs provides an efficient way to generate targeted KI mutations, for example creating equivalent mutations of human-specific variants in rats (Yoshimi et al. 2014) and also in mice (Wu et al. 2013). (4) Finally, because the technology does not rely on using species-specific ES cell lines, it may be possible to adapt it to other mammalian species where it is possible to harvest and manipulate fertilized embryos, such as pigs (Hai et al. 2014; Hauschild et al. 2011), cattle (Tan et al. 2013; Luo et al. 2014), and monkeys (Niu et al. 2014).

Conclusions

The efficient production of genetically modified animals by artificial nucleases, such as ZFNs, TALENs or CRISPR/Cas9, will progress very rapidly. Application of these genome-editing techniques and the use of the huge number of genetically modified rats that are being produced will dramatically accelerate many areas of biomedical research, such as advanced medical studies, drug design, and regenerative medicine.

References

Bassett AR, Tibbit C, Ponting CP, Liu JL (2013) Highly efficient targeted mutagenesis of Drosophila with the CRISPR/Cas9 system. Cell Rep 4:220–228

Bibikova M, Beumer K, Trautman JK, Carroll D (2003) Enhancing gene targeting with designed zinc finger nucleases. Science 300:764

Blaszczyk WM, Arning L, Hoffmann KP, Epplen JT (2005) A Tyrosinase missense mutation causes albinism in the Wistar rat. Pigment Cell Res 18:144–145

Bosma GC, Fried M, Custer RP, Carroll A, Gibson DM, Bosma MJ (1988) Evidence of functional lymphocytes in some (leaky) scid mice. J Exp Med 167:1016–1033

Brown AJ, Fisher DA, Kouranova E, McCoy A, Forbes K, Wu Y, Henry R, Ji D, Chambers A, Warren J, Shu W, Weinstein EJ, Cui X (2013) Whole-rat conditional gene knockout via genome editing. Nat Methods 10:638–640

Capecchi MR (1989) The new mouse genetics: altering the genome by gene targeting. Trends Genet 5:70–76

Cermak T, Doyle EL, Christian M, Wang L, Zhang Y, Schmidt C, Baller JA, Somia NV, Bogdanove AJ, Voytas DF (2011) Efficient design and assembly of custom TALEN and other TAL effector-based constructs for DNA targeting. Nucleic Acids Res 39:e82

Chappel CI, Chappel WR (1983) The discovery and development of the BB rat colony: an animal model of spontaneous diabetes mellitus. Metabolism 32:8–10

Cho SW, Kim S, Kim JM, Kim JS (2013) Targeted genome engineering in human cells with the Cas9 RNA-guided endonuclease. Nat Biotechnol 31:230–232

Christian M, Cermak T, Doyle EL, Schmidt C, Zhang F, Hummel A, Bogdanove AJ, Voytas DF (2010) Targeting DNA double-strand breaks with TAL effector nucleases. Genetics 186: 757–761

Cong L, Ran FA, Cox D, Lin S, Barretto R, Habib N, Hsu PD, Wu X, Jiang W, Marraffini LA, Zhang F (2013) Multiplex genome engineering using CRISPR/Cas systems. Science 339:819–823

Cui X, Ji D, Fisher DA, Wu Y, Briner DM, Weinstein EJ (2011) Targeted integration in rat and mouse embryos with zinc-finger nucleases. Nat Biotechnol 29:64–67

Friedland AE, Tzur YB, Esvelt KM, Colaiacovo MP, Church GM, Calarco JA (2013) Heritable genome editing in *C. elegans* via a CRISPR-Cas9 system. Nat Methods 10:741–743

Fu Y, Foden JA, Khayter C, Maeder ML, Reyon D, Joung JK, Sander JD (2013) High-frequency off-target mutagenesis induced by CRISPR-Cas nucleases in human cells. Nat Biotechnol 31:822–826

Geurts AM, Cost GJ, Freyvert Y, Zeitler B, Miller JC, Choi VM, Jenkins SS, Wood A, Cui X, Meng X, Vincent A, Lam S, Michalkiewicz M, Schilling R, Foeckler J, Kalloway S, Weiler H, Menoret S, Anegon I, Davis GD, Zhang L, Rebar EJ, Gregory PD, Urnov FD, Jacob HJ, Buelow R (2009) Knockout rats via embryo microinjection of zinc-finger nucleases. Science 325:433

Hai T, Teng F, Guo R, Li W, Zhou Q (2014) One-step generation of knockout pigs by zygote injection of CRISPR/Cas system. Cell Res 24:372–375

Hauschild J, Petersen B, Santiago Y, Queisser AL, Carnwath JW, Lucas-Hahn A, Zhang L, Meng X, Gregory PD, Schwinzer R, Cost GJ, Niemann H (2011) Efficient generation of a biallelic knockout in pigs using zinc-finger nucleases. Proc Natl Acad Sci U S A 108:12013–12017

Hockemeyer D, Wang H, Kiani S, Lai CS, Gao Q, Cassady JP, Cost GJ, Zhang L, Santiago Y, Miller JC, Zeitler B, Cherone JM, Meng X, Hinkley SJ, Rebar EJ, Gregory PD, Urnov FD, Jaenisch R (2011) Genetic engineering of human pluripotent cells using TALE nucleases. Nat Biotechnol 29:731–734

Hsu PD, Scott DA, Weinstein JA, Ran FA, Konermann S, Agarwala V, Li Y, Fine EJ, Wu X, Shalem O, Cradick TJ, Marraffini LA, Bao G, Zhang F (2013) DNA targeting specificity of RNA-guided Cas9 nucleases. Nat Biotechnol 31:827–832

Hu X, Chang N, Wang X, Zhou F, Zhou X, Zhu X, Xiong JW (2013) Heritable gene-targeting with gRNA/Cas9 in rats. Cell Res 23:1322–1325

Huang P, Xiao A, Zhou M, Zhu Z, Lin S, Zhang B (2011) Heritable gene targeting in zebrafish using customized TALENs. Nat Biotechnol 29:699–700

Hwang WY, Fu Y, Reyon D, Maeder ML, Tsai SQ, Sander JD, Peterson RT, Yeh JR, Joung JK (2013) Efficient genome editing in zebrafish using a CRISPR-Cas system. Nat Biotechnol 31:227–229

Jacob HJ, Lazar J, Dwinell MR, Moreno C, Geurts AM (2010) Gene targeting in the rat: advances and opportunities. Trends Genet 26:510–518

Jinek M, Chylinski K, Fonfara I, Hauer M, Doudna JA, Charpentier E (2012) A programmable dual-RNA-guided DNA endonuclease in adaptive bacterial immunity. Science 337:816–821

Joung JK, Sander JD (2013) TALENs: a widely applicable technology for targeted genome editing. Nat Rev Mol Cell Biol 14:49–55

Kawano K, Hirashima T, Mori S, Natori T (1994) OLETF (Otsuka Long-Evans Tokushima Fatty) rat: a new NIDDM rat strain. Diabetes Res Clin Pract 24(Suppl):S317–320

Kleinau S, Erlandsson H, Klareskog L (1994) Percutaneous exposure of adjuvant oil causes arthritis in DA rats. Clin Exp Immunol 96:281–284

Lewandoski M (2001) Conditional control of gene expression in the mouse. Nat Rev Genet 2:743–755

Li T, Liu B, Spalding MH, Weeks DP, Yang B (2012) High-efficiency TALEN-based gene editing produces disease-resistant rice. Nat Biotechnol 30:390–392

Li D, Qiu Z, Shao Y, Chen Y, Guan Y, Liu M, Li Y, Gao N, Wang L, Lu X, Zhao Y (2013a) Heritable gene targeting in the mouse and rat using a CRISPR-Cas system. Nat Biotechnol 31:681–683

Li W, Teng F, Li T, Zhou Q (2013b) Simultaneous generation and germline transmission of multiple gene mutations in rat using CRISPR-Cas systems. Nat Biotechnol 31:684–686

Luo J, Song Z, Yu S, Cui D, Wang B, Ding F, Li S, Dai Y, Li N (2014) Efficient generation of myostatin (MSTN) biallelic mutations in cattle using zinc finger nucleases. PLoS One 9:e95225

Ma Y, Shen B, Zhang X, Lu Y, Chen W, Ma J, Huang X, Zhang L (2014a) Heritable multiplex genetic engineering in rats using CRISPR/Cas9. PLoS One 9:e89413
Ma Y, Zhang X, Shen B, Lu Y, Chen W, Ma J, Bai L, Huang X, Zhang L (2014b) Generating rats with conditional alleles using CRISPR/Cas9. Cell Res 24:122–125
Mali P, Aach J, Stranges PB, Esvelt KM, Moosburner M, Kosuri S, Yang L, Church GM (2013a) CAS9 transcriptional activators for target specificity screening and paired nickases for cooperative genome engineering. Nat Biotechnol 31:833–838
Mali P, Yang L, Esvelt KM, Aach J, Guell M, DiCarlo JE, Norville JE, Church GM (2013b) RNA-guided human genome engineering via Cas9. Science 339:823–826
Mashimo T (2014) Gene targeting technologies in rats: zinc finger nucleases, transcription activator-like effector nucleases, and clustered regularly interspaced short palindromic repeats. Dev Growth Differ 56:46–52
Mashimo T, Ohmori I, Ouchida M, Ohno Y, Tsurumi T, Miki T, Wakamori M, Ishihara S, Yoshida T, Takizawa A, Kato M, Hirabayashi M, Sasa M, Mori Y, Serikawa T (2010a) A missense mutation of the gene encoding voltage-dependent sodium channel (Nav1.1) confers susceptibility to febrile seizures in rats. J Neurosci 30:5744–5753
Mashimo T, Takizawa A, Voigt B, Yoshimi K, Hiai H, Kuramoto T, Serikawa T (2010b) Generation of knockout rats with X-linked severe combined immunodeficiency (X-SCID) using zinc-finger nucleases. PLoS One 5:e8870
Mashimo T, Takizawa A, Kobayashi J, Kunihiro Y, Yoshimi K, Ishida S, Tanabe K, Yanagi A, Tachibana A, Hirose J, Yomoda J, Morimoto S, Kuramoto T, Voigt B, Watanabe T, Hiai H, Tateno C, Komatsu K, Serikawa T (2012) Generation and characterization of severe combined immunodeficiency rats. Cell Rep 2:685–694
Mashimo T, Kaneko T, Sakuma T, Kobayashi J, Kunihiro Y, Voigt B, Yamamoto T, Serikawa T (2013) Efficient gene targeting by TAL effector nucleases coinjected with exonucleases in zygotes. Sci Rep 3:1253
Mori M, Hattori A, Sawaki M, Tsuzuki N, Sawada N, Oyamada M, Sugawara N, Enomoto K (1994) The LEC rat: a model for human hepatitis, liver cancer, and much more. Am J Pathol 144:200–204
Mussolino C, Cathomen T (2012) TALE nucleases: tailored genome engineering made easy. Curr Opin Biotechnol 23:644–650
Nakagawa Y, Yamamoto T, Suzuki K, Araki K, Takeda N, Ohmuraya M, Sakuma T (2014) Screening methods to identify TALEN-mediated knockout mice. Exp Anim 63:79–84
Niu Y, Shen B, Cui Y, Chen Y, Wang J, Wang L, Kang Y, Zhao X, Si W, Li W, Xiang AP, Zhou J, Guo X, Bi Y, Si C, Hu B, Dong G, Wang H, Zhou Z, Li T, Tan T, Pu X, Wang F, Ji S, Zhou Q, Huang X, Ji W, Sha J (2014) Generation of gene-modified cynomolgus monkey via Cas9/RNA-mediated gene targeting in one-cell embryos. Cell 156:836–843
Ponce de Leon V, Merillat AM, Tesson L, Anegon I, Hummler E (2014) Generation of TALEN-mediated GRdim knock-in rats by homologous recombination. PLoS One 9:e88146
Porteus MH, Carroll D (2005) Gene targeting using zinc finger nucleases. Nat Biotechnol 23:967–973
Remy S, Tesson L, Menoret S, Usal C, Scharenberg AM, Anegon I (2010) Zinc-finger nucleases: a powerful tool for genetic engineering of animals. Transgenic Res 19:363–371
Sakuma T, Hosoi S, Woltjen K, Suzuki K, Kashiwagi K, Wada H, Ochiai H, Miyamoto T, Kawai N, Sasakura Y, Matsuura S, Okada Y, Kawahara A, Hayashi S, Yamamoto T (2013) Efficient TALEN construction and evaluation methods for human cell and animal applications. Genes Cells 18:315–326
Serikawa T, Yamada J (1986) Epileptic seizures in rats homozygous for two mutations, zitter and tremor. J Hered 77:441–444
Shultz LD, Brehm MA, Garcia-Martinez JV, Greiner DL (2012) Humanized mice for immune system investigation: progress, promise and challenges. Nat Rev Immunol 12:786–798
Stanford WL, Cohn JB, Cordes SP (2001) Gene-trap mutagenesis: past, present and beyond. Nat Rev Genet 2:756–768

Sung YH, Baek IJ, Kim DH, Jeon J, Lee J, Lee K, Jeong D, Kim JS, Lee HW (2013) Knockout mice created by TALEN-mediated gene targeting. Nat Biotechnol 31:23–24

Tan W, Carlson DF, Lancto CA, Garbe JR, Webster DA, Hackett PB, Fahrenkrug SC (2013) Efficient nonmeiotic allele introgression in livestock using custom endonucleases. Proc Natl Acad Sci U S A 110:16526–16531

Tesson L, Usal C, Menoret S, Leung E, Niles BJ, Remy S, Santiago Y, Vincent AI, Meng X, Zhang L, Gregory PD, Anegon I, Cost GJ (2011) Knockout rats generated by embryo microinjection of TALENs. Nat Biotechnol 29:695–696

Tong C, Li P, Wu NL, Yan Y, Ying QL (2010) Production of p53 gene knockout rats by homologous recombination in embryonic stem cells. Nature 467:211–213

Treen N, Yoshida K, Sakuma T, Sasaki H, Kawai N, Yamamoto T, Sasakura Y (2014) Tissue-specific and ubiquitous gene knockouts by TALEN electroporation provide new approaches to investigating gene function in Ciona. Development 141:481–487

Wang H, Yang H, Shivalila CS, Dawlaty MM, Cheng AW, Zhang F, Jaenisch R (2013) One-step generation of mice carrying mutations in multiple genes by CRISPR/Cas-mediated genome engineering. Cell 153:910–918

Wood AJ, Lo TW, Zeitler B, Pickle CS, Ralston EJ, Lee AH, Amora R, Miller JC, Leung E, Meng X, Zhang L, Rebar EJ, Gregory PD, Urnov FD, Meyer BJ (2011) Targeted genome editing across species using ZFNs and TALENs. Science 333:307

Wu J, Kandavelou K, Chandrasegaran S (2007) Custom-designed zinc finger nucleases: what is next? Cell Mol Life Sci 64:2933–2944

Wu Y, Liang D, Wang Y, Bai M, Tang W, Bao S, Yan Z, Li D, Li J (2013) Correction of a genetic disease in mouse via use of CRISPR-Cas9. Cell Stem Cell 13:659–662

Yamamoto S, Nakata M, Sasada R, Ooshima Y, Yano T, Shinozawa T, Tsukimi Y, Takeyama M, Matsumoto Y, Hashimoto T (2012) Derivation of rat embryonic stem cells and generation of protease-activated receptor-2 knockout rats. Transgenic Res 21:743–755

Yamori Y (1991) Overview: studies on spontaneous hypertension-development from animal models toward man. Clin Exp Hypertens A 13:631–644

Yang H, Wang H, Shivalila CS, Cheng AW, Shi L, Jaenisch R (2013) One-step generation of mice carrying reporter and conditional alleles by CRISPR/Cas-mediated genome engineering. Cell 154:1370–1379

Yin H, Xue W, Chen S, Bogorad RL, Benedetti E, Grompe M, Koteliansky V, Sharp PA, Jacks T, Anderson DG (2014) Genome editing with Cas9 in adult mice corrects a disease mutation and phenotype. Nat Biotechnol 32:551–553

Yoshimi K, Tanaka T, Takizawa A, Kato M, Hirabayashi M, Mashimo T, Serikawa T, Kuramoto T (2009) Enhanced colitis-associated colon carcinogenesis in a novel Apc mutant rat. Cancer Sci 100:2022–2027

Yoshimi K, Kaneko T, Voigt B, Mashimo T (2014) Allele-specific genome editing and correction of disease-associated phenotypes in rats using the CRISPR-Cas platform. Nat Commun 5:4240

Chapter 13
Genome Editing in Higher Plants

Yuriko Osakabe and Keishi Osakabe

Abstract Genome editing with engineered nucleases ("GEEN") has emerged as an effective genetic engineering method that uses 'molecular scissors'—artificially engineered nucleases—to digest DNA at targeted locations in the genome of various organisms including plant species. The DNA binding domains of zinc finger (ZF) proteins were first used as plant genome editing tools via the use of designed ZF nucleases (ZFNs), with TAL-effectors (TALE) and the RNA-DNA recognition system CRISPR/Cas9 now being used as powerful genome editing tools to create targeted gene modifications, not only in model plants but also in crop species. The key to genome editing is the introduction of targeted gene-specific double-stranded DNA breaks (DSBs) using the designed endonucleases, then allowing site-directed mutagenesis via nonhomologous end joining (NHEJ) repair and/or gene targeting via homologous recombination (HR), to occur efficiently at specific sites in the genome. This chapter provides an overview of recent advances in genome editing technologies, giving an insight into current plant molecular biology and breeding techniques.

Keywords Breeding • CRISPR/Cas9 • Crop plants • DSB • TALEN • ZFN

13.1 Introduction

Historically, natural mutagenesis has contributed to the development of genetic variation in plant breeding programmes aimed at increasing crop quality and yield. In recent decades the increasing use of induced mutagens has required wide-ranging

Y. Osakabe
RIKEN Center for Sustainable Resource Science,
3-1-1 Koyadai, Tsukuba, Ibaraki 305-0074, Japan

K. Osakabe (✉)
Center for Collaboration among Agriculture, Industry and Commerce, The University of Tokushima, 3-18-15 Kuramoto-cho, Tokushima, Tokushima 770-8503, Japan
e-mail: kosakabe@tokushima-u.ac.jp

T. Yamamoto (ed.), *Targeted Genome Editing Using Site-Specific Nucleases*,
DOI 10.1007/978-4-431-55227-7_13

screening to identify superior traits resulting from mutations that have been introduced randomly into the genome. Radiation is often used for this purpose. Physical and chemical mutagens, such as ethyl methanesulfonate (EMS) and radiation using gamma rays and ion beams, which create DNA damage and consequently induce DNA repair systems, are now well established as means of generating new traits by the random introduction of mutations into the genome. In such mutagenesis schemes, the double-stranded DNA break (DSB) genome repair systems that are programmed in the early processes of meiosis are essential to the generation of mutations in plant species as well as in other organisms (Osakabe et al. 2012). To obtain superior plants using such strategies, screening of the mutagenized populations is required to identify individuals possessing the desired phenotypes.

Genome modifications introduced into crop plants by genetic engineering have been utilized recently to create the potential to improve agricultural practices and add nutritional quality to products. For example, "Golden rice" has been engineered to produce beta-carotene to increase its nutritional food value (Gartland et al. 2013). Silencing of a gene of interest at the posttranscriptional stage using short RNA interference (siRNA) has also been utilized effectively; however, gene knockdown by siRNA can sometimes be variable and incomplete (Fei et al. 2013). On the other hand, sequence-specific modifications have also been widely used in genetic analysis; however, there are some technical difficulties with site-specific mutagenesis in several organisms.

Genome editing with engineered nucleases (GEEN) has been developed as an effective genetic engineering method that uses artificially engineered nucleases to digest DNA at the desired location in the genome of a plant species. An induced DSB introduced at a specific site by the engineered nuclease then undergoes repair by the natural processes of homologous recombination (HR) and nonhomologous end joining (NHEJ). Sequence modifications at the cleaved site, such as deletions or insertions introduced by NHEJ, result in gene disruption and integration of exogenous sequences then occurs in the genome via HR. Currently, three types of engineered nucleases, viz. zinc finger nuclease (ZFN), transcription activator-like effector nuclease (TALEN), and CRISPR (clustered regularly interspaced short palindromic repeats)/Cas9 (CRISPR-associated9), have been developed; of these, TALEN and CRISPR/Cas9 are now especially widely used for genome editing in various plant species (Sander and Joung 2014) (Fig. 13.1).

In this chapter, we describe recent advances in genome editing in the field of plant genome engineering. We highlight genome editing studies that have led to the modification of various plant genomes. Site-directed sequence modifications using engineered nucleases have been used in studies of both model and crop plant species. We also discuss approaches for the application and future prospects of these technologies in plant molecular breeding and biotechnology.

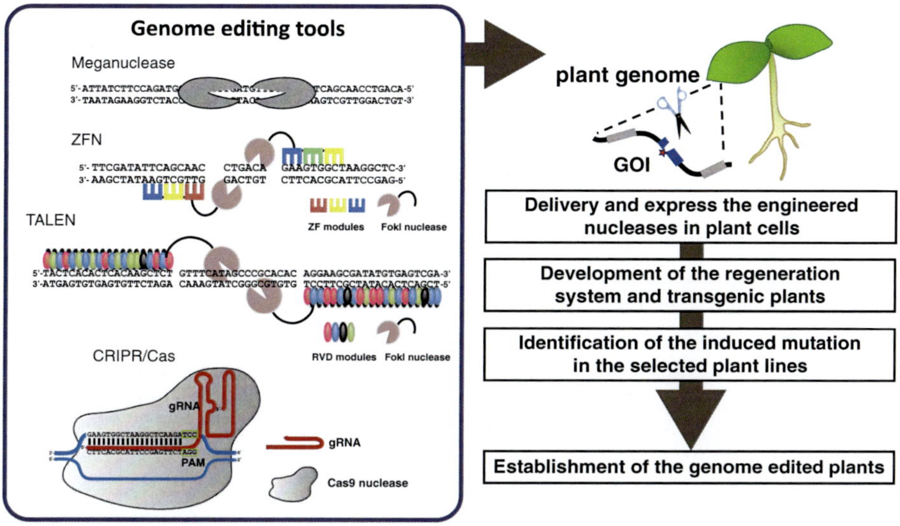

Fig. 13.1 Current methods in GEEN and an overview of the workflow of plant genome editing. Homing endonucleases/meganucleases (EMN) recognize long (~20 base pair) DNA sequences. The *Fok*I nuclease is the DNA-cleavage domain used in ZFN that binds target DNAs via engineered C_2H_2-zinc finger (ZF) domains; TALENs recognize targets via the engineered TALE composed of RVD domains derived from the plant pathogen Xanthomonas. The CRISPR/Cas9 system utilizes RNA-guided engineered nucleases (RGNs), which use a short guide RNA (gRNA) to recognize DNA sequences at the target site

13.2 Site-Directed Mutagenesis of Higher Plants Using Genome Editing Tools

13.2.1 Site-Directed Mutagenesis Using ZFNs

In 2005, Lloyd et al. published the first report of site-directed mutagenesis in plants using ZFN; these authors created a target-specific mutation in *Arabidopsis* using a synthetic model target (Lloyd et al. 2005). Of the ZFN-induced mutations characterized, 78 % were simple deletions of 1–52 bp, 13 % were simple insertions of 1–4 bp, and 8 % were deletions accompanied by insertions. The results showed that the mutants were present in the subsequent generation in 10 % of induced individuals. Using tobacco plants, Wright et al. (2005) also showed ZFN-stimulated gene targeting (GT) in plants. A modular assembly method for ZFN—oligomerized pool engineering (OPEN) to generate ZF arrays that recognize specific DNA sequences—was developed and used for plant genome modification (Townsend et al. 2009). After establishment of the basic assembling systems, site-directed mutagenesis of *Arabidopsis* using ZFN driven by an estrogen-inducible system was reported, resulting in highly effective mutation of the alcohol dehydrogenase and chalcone synthase genes (Zhang et al. 2010). In this latter study, the primary transgenic Arabidopsis induced to express *ADH1* or *TT4* ZFNs exhibited somatic mutation frequencies of 7 % or 16 %, and 69 % or 33 % of the primary lines were transmitted to the next generation (Zhang et al. 2010). We also reported ZFN mutagenesis of the gene *ABSISCIC ACID INSENSITIVE 4* in *Arabidopsis* using the HSP system, and created a new mutant in *abi4* (*ABA-insensitive*) showing the expected response to the plant hormone ABA (Osakabe et al. 2010). Together, looking at these studies and others (Tovkach et al. 2009, 2010), most studies utilizing ZFN have employed the controlled expression of ZFN proteins to create DSBs and induce repair by NHEJ.

13.2.2 Site-Directed Mutagenesis Using TALENs

With further extended methods using TALENs, several studies have succeeded recently in creating target-specific mutations (Cermak et al. 2011; Christian et al. 2013; Gurushidze et al. 2014; Li et al. 2012; Mahfouz et al. 2011; Zhang et al. 2013; Wendt et al. 2013). Voytas' group has developed the assembly of custom TALE arrays in a system known as the Golden Gate cloning method, and achieved site-directed mutation of the *ADH1* gene in *Arabidopsis* protoplasts using this system (Cermak et al. 2011). The de novo engineered TALEN derived from Hax3 from the Brassicaceae pathogen *X. campestris* pv. *armoraciae* strain 5 was used (Mahfouz et al. 2011). Hax3, which recognizes a 12-bp DNA sequence, was used to construct an engineered nuclease that was then used for targeted mutagenesis in a transient assay using *Nicotiana benthamiana* (Mahfouz et al. 2011). Targeted TALEN mutagenesis of the *Arabidopsis* genome using a stable transgenic approach has subsequently been reported (Christian et al. 2013). In this latter study, constitutive TALEN expression in transgenic *Arabidopsis* was induced to create mutations of several genes and a duplicated gene cluster. The frequency of somatic mutagenesis was 41–73 % in the individual transgenic plant lines, and mutants transmitted to the next generation with a frequency of 1.5–12 %. Because of the relatively low toxicity of TALEN compared with ZFN, TALEN can be considered for use in constitutive expressions systems (Christian et al. 2013).

Very recently, TALEN technology has started to be used in crop species such as rice, barley, and maize (Gurushidze et al. 2014; Li et al. 2012; Liang et al. 2014; Shan et al. 2013; Wendt et al. 2013). The rice disease-susceptibility gene and the sucrose-efflux transporter gene *OsSWEET14* were mutagenized by TALEN, resulting in disease-resistant rice with normal phenotypes (Li et al. 2012). Large-scale targeted mutagenesis by TALEN in rice and *Brachypodium* has been reported by Gao's group, who generated knockouts of rice genes and eight *Brachypodium* genes with high efficiency, and also showed large (e.g. 1.3 kb) genomic deletions by simultaneous expression of two pairs of TALENs (Shan et al. 2013). Recently, Gurushidze et al. have developed gene knockout systems based on TALEN in the barley genome using the transformation of embryonic pollen cultures consisting primarily of haploid cells (Gurushidze et al. 2014). These findings have high applicability for detailed studies into gene function and molecular breeding in these various crop species.

13.2.3 Site-Directed Mutagenesis Using CRISPR/Cas9

In the 2 years from the first publication of this system (in 2012) plant genome editing using the CRISPR/Cas9 system was demonstrated in various plant species, such as *Arabidopsis*, tobacco, sorghum, rice, wheat, maize, sweet orange, and liverwort (Feng et al. 2013, 2014; Jiang et al. 2013; Jia and Wang 2014; Li et al. 2013; Liang et al. 2014; Mao et al. 2013; Miao et al. 2013; Nekrasov et al. 2013; Shan et al. 2013; Sugano et al. 2014; Upadhyay et al. 2013), suggesting its highly applicability. The first reports of CRISPR/Cas9 in plant genome editing appeared in August 2013, in a report of transient expression of CRISPR/Cas9 in *Arabidopsis* protoplasts, tobacco cells, and rice plants adapted to create DSBs (Li et al. 2013; Nekrasov et al. 2013; Shan et al. 2013). Li et al., showed that transient expression of a CRISPR/Cas9 vector in *Arabidopsis* protoplasts induced the mutation of homologous members in a multiplex genome as predicted (Li et al. 2013). Using haploid generation of *Marchantia polymorpha* L., Sugano et al., demonstrated a simple and rapid genome editing method employing CRISPR/Cas9 using liverwort as a model species in a study on land plant evolution (Sugano et al. 2014). Recently, Zhu's group reported the multigenerational analysis of CRISPR/Cas9-induced genome editing in *Arabidopsis*, showing the heritable mutation of a 1-bp insertion and short deletion with high efficiency (Feng et al. 2014). In this study, the mutation frequencies were 71.2 % in the first transgenic lines, 58.3 % in the next generation, and 79.4 % in subsequent generations (Feng et al. 2014). Recently, the same group reported efficient gene modification using CRISPR/Cas9 in rice, suggesting that the CRISPR/Cas9 system will become a powerful tool in crop genome engineering (Zhang et al. 2014).

Since the structure of plant genomes and gene families are highly redundant and overlapping, off-target effects during genome editing are an unavoidable and important issue. Recently, Puchta's group reported that applying nickase resulted in efficient genome engineering in *Arabidopsis* (Fauser et al. 2014). Concentrated efforts to extend the above-mentioned findings, together with the precise and detailed evaluation of genome edited-plants to detect off-target effects will be extended to allow further application of genome editing for the crop breeding.

13.3 Gene Targeting and Targeted Gene Addition in Higher Plants Using Genome Editing Tools

Genome editing tools provide novel strategies for genetic manipulation in plants and are likely to assist engineering of desired plant traits by modifying endogenous genes. For instance, the site-specific addition of genes in major crop species can be used for 'trait stacking', whereby several desired traits are physically linked to ensure their co-segregation during the breeding processes. Gene targeting (GT) is a genome engineering method designed to introduce modifications into endogenous genomic sequences via HR. An exogenous DNA with sequences homologous to the target gene and the modification of interest is used as a template instead of undamaged homologous DNA (Osakabe et al. 2006, 2012). HR is induced by creating DSBs at the target site, e.g. the expression of I-Sce I, a rare-cutting restriction enzyme, has been shown to lead to a significant increase in HR-mediated GT in tobacco cells (Puchta et al. 1993, 1996). Rare-cutting enzymes can also lead to the site-specific integration of foreign DNA molecules and induce site-specific mutagenesis.

To introduce GT in both endogenous and exogenous genes with high efficiency in fruit fly and human genomes, engineered ZFNs have been utilized to induce DSB and GT in the presence of donor DNA (Bibikova et al. 2003; Porteus and Baltimore 2003; Urnov et al 2005). ZFNs have also been reported to increase GT frequency in higher plants. A ZFN designed for maize genes and a heterologous donor molecule introduced into maize cells suggested that the ZFN effectively controlled targeted gene addition at a specific site in the genome and that the change was inheritable and transmitted to the next generation (Shukla et al. 2009). Another study by Voytas's group demonstrated ZFN-mediated GT in a transient expression system in tobacco (Townsend et al. 2009). Both studies revealed that over 20 % of selected lines showed GT events. These works demonstrate that cleavage of a chromosomal target by ZFNs dramatically stimulates HR-mediated GT in plants, and provides a basis for future experiments with ZFNs directed to any endogenous genomic location.

Recently, highly efficient targeted gene insertion using TALEN in tobacco protoplasts has also been reported (Zhang et al. 2013). The CRISPR/Cas9 system is also useful in HR-mediated targeted gene insertion in tobacco (Li et al. 2013) and rice (Shan et al. 2013). Since a low HR efficiency has been found in several plant species, appropriate optimizations and improvements will be needed to extend plant genome engineering using custom engineered nuclease-mediated targeting.

13.4 Future Prospects

Current molecular biology techniques allow the direct and effective mutation of particular genes of interest by GEEN. The use of GEEN technology has grown dramatically, the development of CRISPR/Cas9 with simplified methods particularly speeding up the progress of these techniques. Via modification of known genes, these "targetable" nucleases will become a viable alternative to standard breeding methods to identify and introduce novel traits in economically important plants. Genome editing methods now allow us to create newly designed superior traits in various plant species with these applicable and useful methods; however, genome editing will become more effective with further improvements and by conquering several remaining problems such as off-target effects. In particular, a reliable nuclease design, which includes the absence of toxicity and the lack of off-target effects, will be needed for high efficiency and specificity. For example, the double-nicking methods using mutant-types of CAS9 developed by Zhan's group (Ran et al. 2013) and the double cleavage methods of Joung's group (Tsai et al. 2014) will also be utilized in the genome editing of higher plants.

One of the most important issues in plant genome editing is how to deliver and express the engineered nucleases in plant cells, since not all useful plant species are amenable to regeneration and transgenic methods. The appropriate choice of plant tissue and optimized methods for transformation and culturing are still major issues to resolve on the way towards the generation of novel useful crops. Thus, efficient systems for the delivery of genome editing tools into plant cells must be developed. Recent studies using viral vectors, such as recombinant adeno-associated virus (Ellis et al. 2013) and geminivirus-based replicons (Baltes et al. 2014), to deliver the DNA of genome editing tools has enabled efficient genome engineering in various plant species. As new plant breeding techniques, these efforts, together with a deeper understanding of whole genome structure and function, promise to deliver future technologies in breeding new and important traits in various plants.

Acknowledgments The authors are grateful to Dr. S. Toki for his kind support to this work. This work was supported by the Grant-in-Aid for Scientific Research C from the Japan Society for the Promotion of Science (21580125 to Y.O. and 25450002 to K.O.).

References

Baltes NJ, Gil-Humanes J, Cermak T, Atkins PA, Voytas DF (2014) DNA replicons for plant genome engineering. Plant Cell 26:151–163

Bibikova M, Beumer K, Trautman JK, Carroll D (2003) Enhancing gene targeting with designed zinc finger nucleases. Science 300:764

Cermak T, Doyle EL, Christian M, Wang L, Zhang Y, Schmidt C, Baller JA, Somia NV, Bogdanove AJ, Voytas DF (2011) Efficient design and assembly of custom TALEN and other TAL effector-based constructs for DNA targeting. Nucleic Acids Res 39:e82

Christian M, Qi Y, Zhang Y, Voytas DF (2013) Targeted mutagenesis of *Arabidopsis thaliana* using engineered TAL effector nucleases. G3 (Bethesda) 3:1697–1705

Ellis BL, Hirsch ML, Porter SN, Samulski RJ, Porteus MH (2013) Zinc-finger nuclease-mediated gene correction using single AAV vector transduction and enhancement by Food and Drug Administration-approved drugs. Gene Ther 20:35–42

Fauser F, Schiml S, Puchta H (2014) Both CRISPR/Cas-based nucleases and nickases can be used efficiently for genome engineering in *Arabidopsis thaliana*. Plant J 79:348–359

Fei Q, Xia R, Meyers BC (2013) Phased, secondary, small interfering RNAs in posttranscriptional regulatory networks. Plant Cell 25:2400–2415

Feng Z, Zhang B, Ding W, Liu X, Yang DL, Wei P, Cao F, Zhu S, Zhang F, Mao Y, Zhu JK (2013) Efficient genome editing in plants using a CRISPR/Cas9 system. Cell Res 23:1229–1232

Feng Z, Mao Y, Xu N, Zhang B, Wei P, Yang DL, Wang Z, Zhang Z, Zheng R, Yang L, Zeng L, Liu X, Zhu JK (2014) Multigeneration analysis reveals the inheritance, specificity, and patterns of CRISPR/Cas9-induced gene modifications in Arabidopsis. Proc Natl Acad Sci U S A 111:4632–4637

Gartland KM, Bruschi F, Dundar M, Gahan PB, Viola Magni M, Akbarova Y (2013) Progress towards the 'Golden Age' of biotechnology. Curr Opin Biotechnol 24(Suppl 1):S6–S13

Gurushidze M, Hensel G, Hiekel S, Schedel S, Valkov V, Kumlehn J (2014) True-breeding targeted gene knock-out in barley using designer TALE-nuclease in haploid cells. PLoS One 9:e92046

Jia H, Wang N (2014) Targeted genome editing of sweet orange using Cas9/sgRNA. PLoS One 9:e93806

Jiang W, Zhou H, Bi H, Fromm M, Yang B, Weeks DP (2013) Demonstration of CRISPR/Cas9/sgRNA-mediated targeted gene modification in Arabidopsis, tobacco, sorghum and rice. Nucleic Acids Res 41:e188

Li T, Liu B, Spalding MH, Weeks DP, Yang B (2012) High-efficiency TALEN-based gene editing produces disease-resistant rice. Nat Biotechnol 30:390–392

Li JF, Norville JE, Aach J, McCormack M, Zhang D et al (2013) Multiplex and homologous recombination–mediated genome editing in Arabidopsis and *Nicotiana benthamiana* using guide RNA and Cas9. Nat Biotechnol 31:688–691

Liang Z, Zhang K, Chen K, Gao C (2014) Targeted mutagenesis in *Zea mays* using TALENs and the CRISPR/Cas9 system. J Genet Genomics 41:63–68

Lloyd A, Plaisier CL, Carroll D, Drews GN (2005) Targeted mutagenesis using zinc-finger nucleases in Arabidopsis. Proc Natl Acad Sci U S A 102:2232–2237

Mahfouz MM, Li L, Shamimuzzaman M, Wibowo A, Fang X, Zhu JK (2011) De novo-engineered transcription activator-like effector (TALE) hybrid nuclease with novel DNA binding specificity creates double-strand breaks. Proc Natl Acad Sci U S A 108:2623–2628

Mao Y, Zhang H, Xu N, Zhang B, Gou F, Zhu JK (2013) Application of the CRISPR/Cas9 system for efficient genome engineering in plants. Mol Plant 6:2008–2011

Miao J, Guo D, Zhang J, Huang Q, Qin G, Zhang X, Wan J, Gu H, Qu LJ (2013) Targeted mutagenesis in rice using CRISPR/Cas9 system. Cell Res 23:1233–1236

Nekrasov V, Staskawicz B, Weigel D, Jones JDJ, Kamoun S (2013) Targeted mutagenesis in the model plant Nicotiana benthamiana using Cas9 RNAguided endonuclease. Nat Biotechnol 31:691–693

Osakabe K, Abe K, Yoshioka T, Osakabe Y, Todoriki S, Ichikawa H, Hohn B, Toki S (2006) Isolation and characterization of the RAD54 gene from *Arabidopsis thaliana*. Plant J 48:827–842

Osakabe K, Osakabe Y, Toki S (2010) Site-directed mutagenesis in Arabidopsis using custom-designed zinc finger nucleases. Proc Natl Acad Sci U S A 107:12034–12039

Osakabe K, Saika H, Okuzaki A, Toki S (2012) Site-directed mutagenesis in higher plants. In: Shu QY, Forster BP, Nakagawa H (eds) Plant mutation breeding and biotechnology. CAB eBooks, Wallingford, pp 523–533

Porteus MH, Baltimore D (2003) Chimeric nucleases stimulate gene targeting in human cells. Science 300:763

Puchta H, Dujon B, Hohn B (1993) Homologous recombination in plant cells is enhanced by in vivo induction of double strand breaks into DNA by a site-specific endonuclease. Nucleic Acids Res 21:5034–5040

Puchta H, Dujon B, Hohn B (1996) Two different but related mechanisms are used in plants for the repair of genomic double-strand breaks by homologous recombination. Proc Natl Acad Sci U S A 93:5055–5060

Ran FA, Hsu PD, Lin CY, Gootenberg JS, Konermann S, Trevino AE, Scott DA, Inoue A, Matoba S, Zhang Y, Zhang F (2013) Double nicking by RNA-guided CRISPR/Cas9 for enhanced genome editing specificity. Cell 154:1380–1389

Sander JD, Joung JK (2014) CRISPR/Cas9 systems for editing, regulating and targeting genomes. Nat Biotechnol 32:347–355

Shan Q, Wang Y, Li J, Zhang Y, Chen K et al (2013) Targeted genome modification of crop plants using a CRISPR/Cas9 system. Nat Biotechnol 31:686–688

Shukla VK, Doyon Y, Miller JC, DeKelver RC, Moehle EA, Worden SE, Mitchell JC, Arnold NL, Gopalan S, Meng X, Choi VM, Rock JM, Wu YY, Katibah GE, Zhifang G, McCaskill D, Simpson MA, Blakeslee B, Greenwalt SA, Butler HJ, Hinkley SJ, Zhang L, Rebar EJ, Gregory PD, Urnov FD (2009) Precise genome modification in the crop species Zea mays using zinc-finger nucleases. Nature 459:437–441

Sugano SS, Shirakawa M, Takagi J, Matsuda Y, Shimada T, Hara-Nishimura I, Kohchi T (2014) CRISPR/Cas9-mediated targeted mutagenesis in the liverwort Marchantia polymorpha L. Plant Cell Physiol 55:475–481

Tovkach A, Zeevi V, Tzfira T (2009) A toolbox and procedural notes for characterizing novel zinc finger nucleases for genome editing in plant cells. Plant J 57:747–757

Tovkach A, Zeevi V, Tzfira T (2010) Validation and expression of zinc finger nucleases in plant cells. Methods Mol Biol 649:315–336

Townsend JA, Wright DA, Winfrey RJ, Fu F, Maeder ML et al (2009) High-frequency modification of plant genes using engineered zinc-finger nucleases. Nature 459:442–445

Tsai SQ, Wyvekens N, Khayter C, Foden JA, Thapar V, Reyon D, Goodwin MJ, Aryee MJ, Joung JK (2014) Dimeric CRISPR RNA-guided FokI nucleases for highly specific genome editing. Nat Biotechnol 32:569–576

Upadhyay SK, Kumar J, Alok A, Tuli R (2013) RNA-guided genome editing for target gene mutations in wheat. G3 (Bethesda) 3:2233–2238

Urnov FD, Miller JC, Lee YL, Beausejour CM, Rock JM, Augustus S, Jamieson AC, Porteus MH, Gregory PD, Holmes MC (2005) Highly efficient endogenous human gene correction using designed zinc-finger nucleases. Nature 435:646–651

Wendt T, Holm PB, Starker CG, Christian M, Voytas DF, Brinch-Pedersen H, Holme IB (2013) TAL effector nucleases induce mutations at a pre-selected location in the genome of primary barley transformants. Plant Mol Biol 83:279–285

Wright DA, Townsend JA, Winfrey RJ Jr, Irwin PA, Rajagopal J, Lonosky PM, Hall BD, Jondle MD, Voytas DF (2005) High-frequency homologous recombination in plants mediated by zinc-finger nucleases. Plant J 44:693–705

Zhang F, Maeder ML, Unger-Wallace E, Hoshaw JP, Reyon D et al (2010) High frequency targeted mutagenesis in *Arabidopsis thaliana* using zinc finger nucleases. Proc Natl Acad Sci U S A 107:12028–12033

Zhang Y, Zhang F, Li X, Baller JA, Qi Y, Starker CG, Bogdanove AJ, Voytas DF (2013) Transcription activator-like effector nucleases enable efficient plant genome engineering. Plant Physiol 161:20–27

Zhang H, Zhang J, Wei P, Zhang B, Gou F, Feng Z, Mao Y, Yang L, Zhang H, Xu N, Zhu JK (2014) The CRISPR/Cas9 system produces specific and homozygous targeted gene editing in rice in one generation. Plant Biotechnol J 12:797–807

Kolibri 360 GmbH
Ferdinand-Jühlke-Straße 7
99095 Erfurt, Deutschland
produktsicherheit@kolibri360.de